PROFESSION

CW01506747

CITIZEN MANUAL 5
LAND NAVIGATION

Developed by
The Professional Citizen Project

www.TPCproject.com

For my Mom and Dad

Thank you for everything you did for me. All your sacrifices that set the conditions to make me the person I am today. You were the best parents a man could ever ask for. I miss you both very much.

Citizen Manual 5 (CM-5)
Land Navigation

ISBN: 979-8-9895092-4-9

Printed in the USA

www.TPCproject.com

Contents   Page

Introduction   1

Navigation Process   6
Navigation Methods   7

Chapter 1 – Navigation Tools   9

Maps   11
Compasses   40
Scales / Protractors   48
GPS   51
Navigation Kit   59
LRF, Pace Beads, Altimeter, Drones   58

Chapter 2 – Know Where You Are   61

Grid Locations   62
Determine a Grid Coordinate   63
Terrain Features   69
Elevation and Relief   78
Slope   87

Contents   Page

Chapter 3 –Plan the Route   90

METT-TC Review   91
OAKOC   95
Route Selection   103
Nav Offset   108
Routes During a Crisis   112

Chapter 4 – Navigate / Stay on the Route   116

Direction   117
Baselines   121
Grid Azimuth (GAZ)   124
Azimuth Conversion   125
Use a Protractor   131
Measure Grid Azimuth   133
Measure Distance   138

Chapter 5 – Mounted Land Navigation   180

Mounted Terrain Association   189
Bypasses   193
Street Navigation Terms   194
Interstate System   196
Reverse Planning   199

APPENDICES

APPENDIX A - Land Nav Gear Checklist   203
APPENDIX B - USGS Map Legend   204
APPENDIX C - Map Modifications   207
APPENDIX D - Reverse Planning Timeline   208
APPENDIX E - Pace Count Record   209

References   210

"I can't say I was ever lost, but I was bewildered once for three days,"

- David Crockett

INTRODUCTION

There is something primitive about it. The sense of pride and the visceral (albeit quiet and internal) reaction when you step out of the woodline after a long movement and your objective is exactly where you knew it would be. It may be an orange and white control marker, the trail head parking lot where you left your truck, or an obscure piece of defensible ground where you set your patrol in for the night. The respect for those who can successfully find their way across the terrain is as old as mankind itself. Military professionals pride themselves on it; those who cannot nav are in the "I would never follow him" category. It takes study and a lot of on the ground practice - but anyone can become an expert at this. Becoming a proficient navigator is founded in process, attention to detail, and confidence in your abilities built over time. The confidence gained by successfully navigating the dark woods alone at night with only a compass, map, and protractor is seldom matched.

Land navigation is a complex skillset, a perfect blend of art and science. Start with basic map reading: locate a point on a map by grid, identify terrain features on a map, and navigate by dead reckoning during both day and night. Enhance these skills by diving into terrain association, terrain analysis, detailed route planning, intersection, resection, and other land nav skills. Land navigation takes practical application to build proficiency. You will not learn it by only reading this manual; it takes a lot of time in the field.

This Manual, CM-5, Land Navigation

This manual provides you with land navigation technical expertise and best practices learned through years of experience. Our approach for The Project is always realistic and delivered to you in the context of the "typical" American dad, mom, daughter, or son. This is written in a straightforward, no fluff manner. We focus on the realistic navigation skills that will apply to conditions you are likely to encounter. This is a practical, high odds use case manual for the normal human. We built this for you to use as a foundation reference, it is detailed and straightforward.

Some elements you may have been taught in large format land nav classes are not in here. **This is not a comprehensive manual by any means, there are hundreds of additional pages that can be written about land navigation.** This manual is laser focused on the foundational land nav skills you need in the real world.

The progression of subjects in the chapters may not be the same as what you have seen elsewhere, but do not skip ahead. The sequence of this manual is a proven method for modern adult learners.

The skills in this manual nest with the other books in the Professional Citizen series. You will find land nav info in the other CM's; some of that will be reiterated in here out of necessity. We must include portions of that content from the previous books to allow this manual to stand on its own. This is a holistic, thorough primary reference you can use to learn and teach others.

THE FRAMEWORK

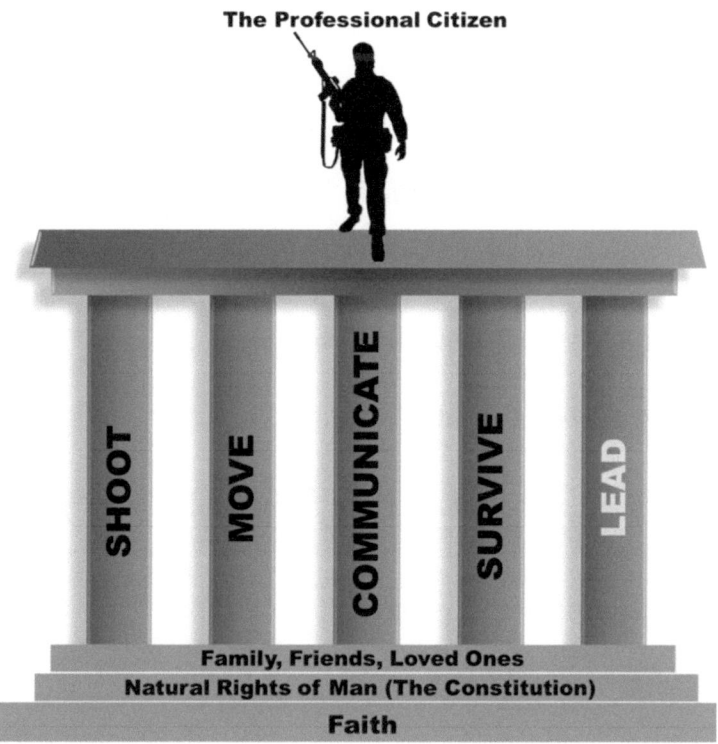

The Professional Citizen

SHOOT   MOVE   COMMUNICATE   SURVIVE   LEAD

Family, Friends, Loved Ones
Natural Rights of Man (The Constitution)
Faith

The Professional Citizen must be able to **Shoot, Move, Communicate, Survive,** and **Lead.** This manual (the CM-5) is focused squarely on the move skillset by providing detailed content on the land navigation subject. This manual is an expansion on the introductory Land Nav chapter in the Individual Tactical Skills Manual (CM-1).

HOW THIS MANUAL IS ORGANIZED

This manual is aligned with the navigation steps. We will address components, tools, and techniques in the logical order you will need for a task or mission that requires land navigation skills.

Navigation Tools (Chapter 1)

This section describes in detail the tools we use for navigation. Maps, compasses, GPS units and other items that help us navigate.

Know Where You Are (Chapter 2)

Covers the skills to accurately spot and plot your location on a map. Understanding grid reference systems and how they work in conjunction with the terrain.

Plan a Route (Chapter 3)

This chapter covers the blend of art and science of visualizing and planning your route through the terrain.

Navigate the Route and Recognize the Objective (Chapter 4)

Chapter 4 covers the skills that keep you on track while navigating along a route. This process concludes with the final step of recognizing the objective. (the last two steps of the navigation process are combined in this one chapter)

Mounted Land Nav (Chapter 5)

In this brief overview chapter we will discuss some of the additional requirements and nuances of navigating while mounted (in or on a vehicle).

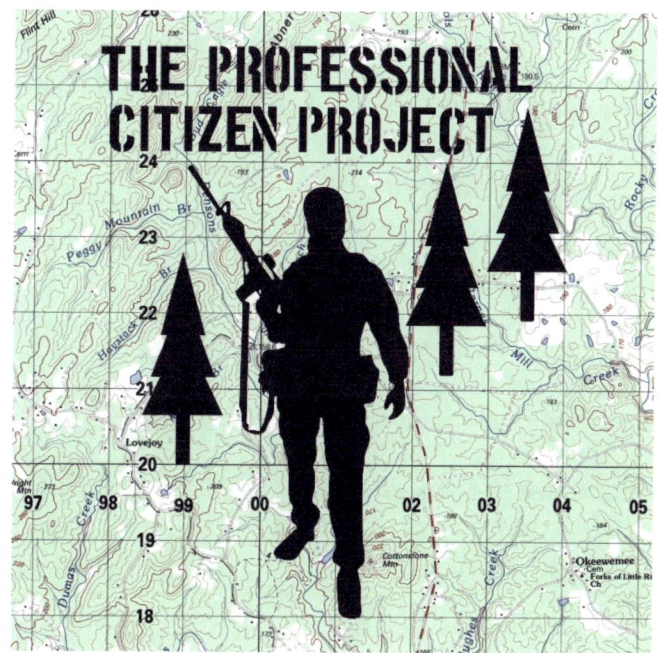

NAVIGATION

We don't just navigate for navigation's sake (unless it is recreational or for navigation training). Always keep in mind you are using the skill to support another task or mission. You might be traveling to a vacation spot or navigating through the wilderness as part of a search and rescue operation to save someone's life. The consequences of delays or missing your objective due to a navigation error differ for these, but the process to get where you want to go are the same.

THE PROCESS

Navigating from one place to another has four components or fundamentals. These are:

Know your current location or your start point. Knowing where you are in relation to the terrain and the map.

Plan your route. Analyzing terrain to choose the best route to an objective.

Execute / follow your route. The mechanics and skills to move along a route as well as the adjustments and navigational challenges along the way.

Recognize the objective. Arriving at and recognizing the objective when you get there.

METHODS OF NAVIGATING

There are several methods of land navigation. The best approach is to use the applicable pieces from each of these and apply your mission or navigation task. Use whichever method or combination that best suits your situation.

Terrain Association

Terrain association is matching up the terrain features on the ground with the contour intervals on your map. As you move you will use the terrain to keep yourself oriented by matching them up with your map.

General Azimuth

This is when you follow or guide off a linear terrain feature or set of features along a general direction. Traveling north along a valley is a good example of this. As you can immediately see, this method has tactical disadvantages since following or handrailing a ridge or valley will often put you along a natural line of drift (more on this later). An advantage of this method is it speeds movement and simplifies navigation.

Dead Reckoning (Point Navigation)

Dead reckoning is starting from a known point and using a distance and direction. This method uses a compass and a distance from the pace man (or a vehicle's odometer when mounted) to follow a direct route or leg.

GPS Waypoint

This method is like the familiar smartphone GPS apps. The practical difference is the waypoints your app uses along a route are auto generated. Handheld GPS navigation for dismounted land nav is best done by entering separate locations (waypoints) that builds "legs" to manage a realistic navigation route. The GPS directs you to the selected end point or waypoint by means of electronic dead reckoning (for non-mapping GPS units). The greatest benefit of GPS navigation is the system is in a constant state of course correction. With a good signal the device will continually help you stay on course.

Combination

Combining parts of each of these navigation methods is the way to go. Terrain association and general azimuth method enable you to set a rough compass bearing and move as quickly as the situation allows toward a catching feature or a navigational attack point. Once reached, you will switch to point navigation by paying close attention to detail, taking as much time as necessary to analyze the situation and find the objective or point. Terrain association and general azimuth method allow for some flexibility in the movement and do not require the same level of control as point navigation (aka dead reckoning).

CHAPTER 1
NAVIGATION TOOLS

Knowledge and skills are more important than gear, but this isn't as clean or absolute as it sounds. We still need high quality equipment to be effective, there is no amount of skill you can develop to consistently determine where magnetic north is without a compass. And unless you are blessed with a photographic memory the need for a map to guide your movement is necessary. Understanding and mastering the tools of land navigation -- map, compass, and protractor are necessary for mission success. This section will describe and discuss the details of the navigation tools available to us.

THE BASIC TOOLS

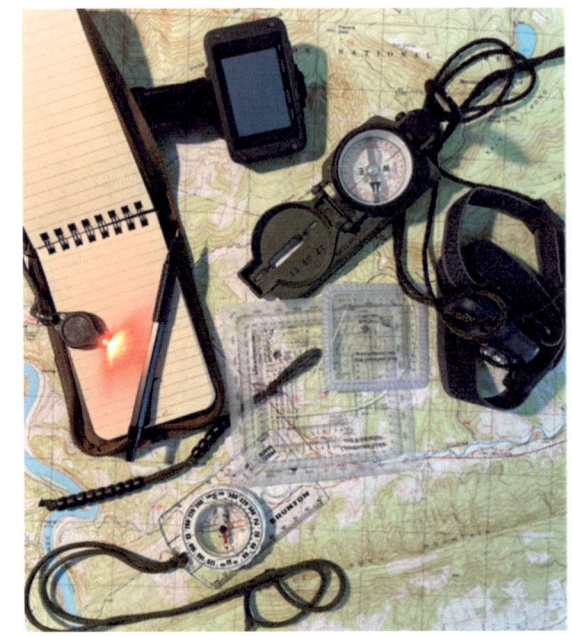

- Map
- Compass
- Protractor
- Pencil (.7 to .9 mm mechanical works best)
- Superfine Map Marker
- Waterproofing/Map Case
- Pace Beads
- Notebook
- Straight Edge (can be the protractor)

You need a high-quality set of navigation tools with redundant options. Do not cut corners when selecting your nav equipment; weatherproof, durable, and proven are the requirements for these items.

MAPS

A map is a scaled graphic representation of a portion of the earth's surface as seen from above. It uses colors, symbols, and labels to represent ground features. To keep the discussion relevant, we will address the three common maps used for ground navigation. These are a standard road map, a specific area map, (both of these are types of *planimetric* maps), and the topographic map. Topographic map use is the main focus of this manual.

Planimetric Maps show the positions of features without showing their relationship to the terrain or elevation. Examples include highways, rivers, lakes, and other manmade features the mapmaker wished to convey. These types of maps include road maps, road atlases, and specific area maps like you would use at an amusement park. These maps are typically NTS or *not to scale*.

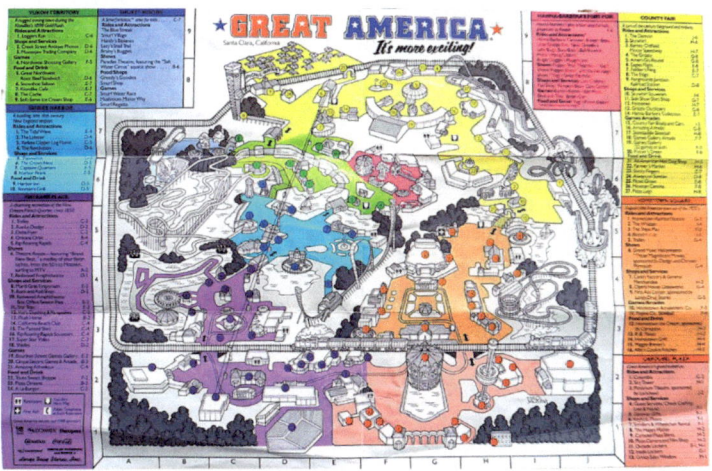

A specific area map (a type of planimetric map) of an amusement park.

Topographic maps differ from planimetric maps by adding details of vertical positions of features (relief). These topographic (or topo) maps are what we should use for precision land navigation.

A topographic map provides information on the existence, location, and distance between ground features. Topo maps show variations in terrain, heights of natural features, and vegetation cover (to a degree). Topo maps portray terrain features as well as the horizontal positions of the features represented. The relief of features is represented by contour lines or shading. The elevations and contours are measured from a specific vertical datum plane, usually mean sea level. The height of the features, relief is always referred to as elevation, not altitude when we are discussing land navigation.

Topographic maps are the primary map for dismounted land navigation. They provide a large amount of information in a small package.

SCALE

Scale is the ratio or relationship a drawn feature on a map has to its actual size on the ground. Most maps we use have representative fraction (RF) scales. These are a ratio of map to ground measurement written with numbers on each side of a colon. The first number is always a "1" and the second is the number of units on the ground. A 1:24,000 means every unit on the map equals 24,000 units on the ground. Sometimes you will hear scales said as "one to 24" or "1:50".

The larger the second number, the larger the ratio and the smaller the map scale. The *small scale* maps lose detail compared to *large scale* maps. The cartographer must draw more features in a smaller space on small scale maps. For example, a 1:24,000 map will have greater detail than a 1:50,000.

Size and scale comparison of approximately the same map area for a 1:24000 map on the left and a 1:50000 topo map on the right. This example demonstrates the level of detail and the actual size between these two scales. The advantage in detail of the 1:24 is readily apparent.

There are numerous map scales in use, but we usually find ourselves with one of these common / standard topographic map scales below:

1:24,000 - usually found on 7.5 minute USGS topo maps, excellent scale for dismounted land nav.

1:25,000 - some military maps and select USGS. Another good scale to use for dismounted nav.

1:50,000 - standard for MGRS military maps, good balance for mounted and dismounted land nav use.

1:100,000 - less common but excellent for large areas of operation and fast moving mounted land nav.

The detail in the Delorme/Garmin bound topo state maps is excellent. The scales vary for different map sets though, so dial your setup in well ahead of time.

14

CHOOSE THE RIGHT MAP FOR THE JOB

Deciding which maps you need is not as straightforward as it seems and it can be a challenge. Maps are not expensive individually, but when you build a collection of map sheets to cover an extended area the cost adds up. Do you need a set of 7.5 minute 1:24,000 paper topo maps for your entire state? Probably not. Conversely, we can't rely on a single small scale state road atlas to terrain associate or report grid coordinates in our local area.

Assess your AO (Area of Operations) and Area of Interest (AI) to determine how much map detail you need. Try using the next two graphics as a guide to generate thought and help you decide. The AO and AI specific definitions / determinations are beyond the scope of this manual, but this simplified version may give you an idea of the maps you should acquire.

Build your map sets to account for escape routes from natural disasters, routes to linkup with loved ones that may be away at school, or alternate routes back home if you get caught at work during a disaster. Pick your maps to meet specific requirements, not just focused on the 1:50,000 topo map that comes to mind when you think of "land nav." Dismounted property and neighborhood patrols during a disaster will require high-resolution large-scale maps, but the flood evac route for your family will require a smaller scale road map. Be sure to plan for all of these.

Visualize and identify the areas where you need map coverage. Consider the concepts in the next two diagrams to help you decide what you need:

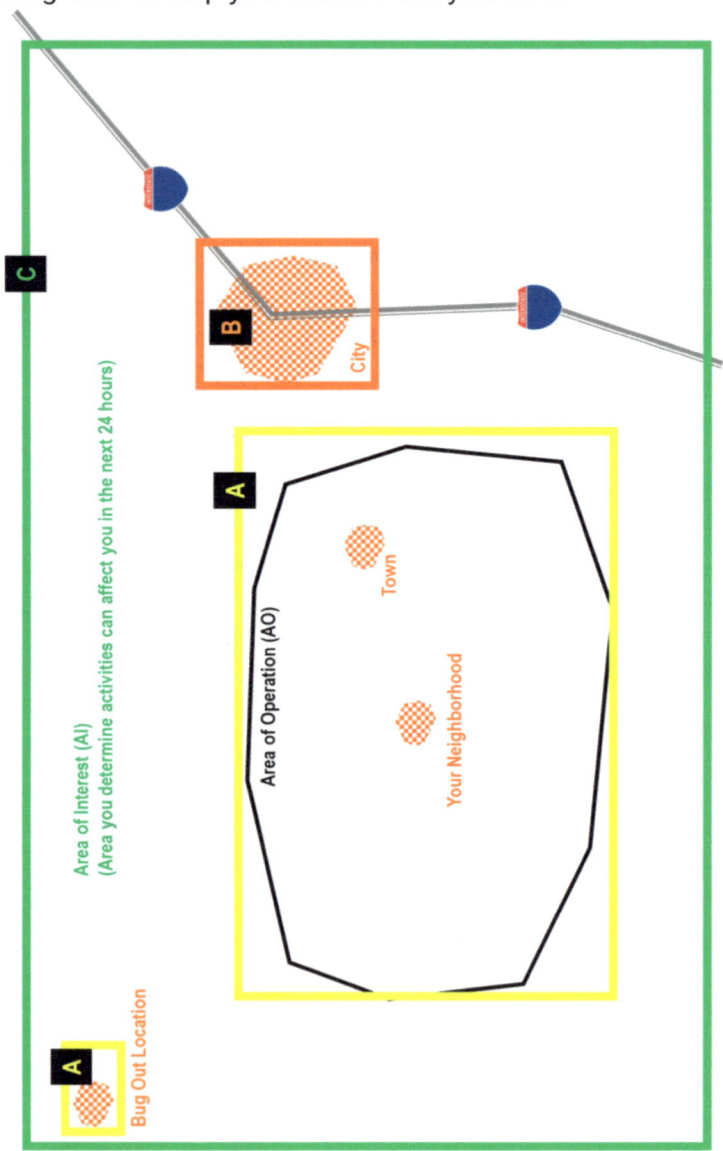

Area of Interest (AI)
(Area you determine activities can affect you in the next 24 hours)

C

B

City

A

Area of Operation (AO)

Town

Your Neighborhood

A

Bug Out Location

A The area you predict you will need the highest resolution of terrain data.

Maps needed:
    -Large scale topographic maps (1:24,000 or 1:50,000)
        -Satellite Imagery
        -Orthoimage

B Large Urban Areas in your area or along expected routes.

Maps needed:
        -City street map (city specific, not the insets or expanded maps on a state map)
        -State road map
        -Large scale (1:24000 or 1:25000 topographic map

C The expanded area that may influence your actions or you may have to travel through.

Maps needed:
        -Small scale topo map (1:100,000). These may be difficult to find, so don't get too wrapped up over it
        -Set of small / medium scale topo maps (state books are readily available)
        -State Road Map
        -Road atlas to cover areas that "feed" into your AO
        * you may need high resolution maps of select areas you will travel through, for example along your bug out route(s)

Once you write additional information like routes, operations graphics, checkpoints etc. on your map it becomes a sensitive item. In the context of a widespread crisis if a marked map falls into unauthorized hands, it could provide information to a threat. Even more risky would be a map that has friendly movements and positions marked that a criminal element can use. Ensure you have a destruction plan in place for these items, protecting this info is critical to security for your team.

17

SOURCES FOR PRINTED MAPS

Now that you have decided which maps you need, where can you get them? Electronic maps are easy to acquire, but you should also have current sets of paper topo maps on hand. Your ability to download an electronic map after a crisis may be limited since networks may be disabled or interrupted.

We can source road maps, atlases, and even satellite imagery easily through the current commercial market. Big box stores and online warehouses have state road maps and state sets of smaller scale topo maps (like the Delorme™ state topo atlases). Remember to rotate your maps out as they get old, depending on the rate of construction and change in your area a 3 to 5 year replacement is a good rule of thumb for road maps.

Finding large scale topographic maps can be a challenge if you have never done so. Outfitters and outdoor stores will have topo maps, but these tend to be focused on nearby recreation areas or national forests. The large scale 1:24,000 that includes the area around your neighborhood may be difficult to source locally. Fortunately, there are some great options to get what we need. Many of these sources provide downloadable maps (some are free) with the ability to print at home which is perfect for small area maps. Our focus is acquiring professionally printed maps for this section of the manual. We will show you a few sources, you can choose your own of course.

The sites below are not an endorsement or sponsored, these are only presented as sources for consideration:

USGS Topo Maps
https://store.usgs.gov/#TM

US Forest Service
https://data.fs.usda.gov/geodata/rastergateway/states-regions/states.php

CalTopo
https://caltopo.com/map.html

MyTopo
https://mapstore.mytopo.com/

NOT ALL MAPS ARE CREATED EQUAL

The current age of topographic maps and cartography methods is not as good as we would like it to be. The map data that most government and commercial mapmakers (map printers to be more accurate) is digital data sourced from The National Map. If you look at an older USGS or Defense Mapping Agency topo map you immediately notice the detail is impressive. It was an art form to map an area - but it took years to produce an updated version of a topo map. Surveys, walking the ground, and physically verifying the features was the way these older maps were done. It was slow and resource intensive, but the resulting product was phenomenal.

The benefit we have today is the digital map data is no more than three years old. USGS Topo maps are updated on a three-year cycle with one third of the country updated each year. This does not include topos for Alaska which are on a different schedule, but Alaskan mapping has always been the outlier.

The features shown on the new maps are generated from the latest data in The National Map and other standard sources. The USGS no longer does field verification or other primary data collection for these feature classes, and *there are no national data sources suitable for general-purpose, 1:24,000-scale maps*. This is what may surprise some of you - especially if you are accustomed to using "old school" topo maps. If it existed on the ground, the mapmaker found a way to add it to our maps.

Some features we were accustomed to on our topo maps did not transfer to the newer digital processes. There are some significant things missing from new map data that expert navigators grew up using. Examples of these now unmarked features can include recreational trails, pipelines, power lines, survey markers, and many types of buildings. Since US Topo maps are mass produced from national databases, some features shown on traditional maps might never be included on future US Topo maps. For example, a national database of isolated ranch windmills and water tanks (which were on the manually built maps) is unlikely to ever be built.

The photos above are of the same terrain, the same scale, the same grid square. On the left is a USGS "historical" map. You can see the details that are left off the new US Topo print on the right. No buildings, churches, power lines, or spot elevations. The power line running NW / SE is a great catching feature, but using the new map you would never know it was there until you walked up on it. To mitigate this loss of detail we must use other tools like satellite imagery and aerial photos to augment the new version topo maps.

This isn't a complaint or "the world has gone to hell" commentary on cartography, I just want to point out some things have changed on our maps. The 60 year old power line cutting through the forest may not show up on your two year old map - don't let it throw you when navigating a remote area. Studying your intended area using satellite and aerial photos in conjunction with your map is always advised. Using a historical series map with all the features (USGS still sells these) to augment modern topos can be beneficial, just pay attention to the datum version (more on this later).

Traditional topo maps labeled many public buildings and structures such as courthouses, libraries, transportation terminals, and bridges. National public domain digital datasets of these feature classes do not exist, so many are not printed on new topo maps.

Powerlines, oil and gas pipelines, and other energy infrastructure are not shown on new US Topo maps. Digital datasets do not exist (and USGS claims there are security reasons for not publishing these data). You will need to rely on historical topo maps and overhead imagery for the location of these features.

Unique landmark features such as buildings, natural features, isolated monuments, and points of interest are not in the national GIS database. Recreational features such as campgrounds, boat docks, golf courses, etc will never be a high priority for USGS mapping so they may or may not appear on your map.

Forest Service Topo Maps. USFS topo maps (FSTopos) use the same format as the USGS, but with enhancements and regular revisions contributed by USFS field staff that emphasize forest-related information. These maps only cover US Forest Service areas. Excellent trail coverage and prominently marked Forest Service Road numbers on these are helpful for navigating backcountry areas; they also accentuate trails, roads, and water features. The identifying routes and trails make these maps valuable for outdoor recreation and especially useful when training new or younger navigators.

Game Management Unit (GMU) Maps. These are great maps to use for mounted land nav in the western US. They are 1:100,000 base maps with Game Management Unit (GMU) boundaries as an overlay. The colors can get a bit out of control and distracting, but as a baseline 1:100 with contour lines and accurate features they are still a good option.

ORTHOIMAGE

Due to fewer details on new series topo maps we need as much additional info as we can get. The "historical" series of maps from USGS and even old copies of topo maps you can source locally will help, but we need up to date imagery to complement current series maps. This is especially true for built up (urban) areas. A digital orthophoto quadrangle (DOQ)--or any orthoimage--is **a computer-generated image of an aerial photograph in which displacements (distortions) caused by terrain relief and camera tilts have been removed.** It combines the image characteristics of a photograph with the geometric qualities of a map. Unlike an aerial photograph, an orthoimage has a uniform scale, so it can be used as a base map (if printed correctly to maintain scale). It is possible to measure directly on an orthoimage, just like other maps.

An example of an orthoimage from the USGS website. These can be tailored to your needs and matched to your map sheet(s). They are a valuable addition to current topo maps.

The High Resolution Orthoimagery collection was acquired through USGS contracts, partnerships with other Federal, state, tribal, or regional agencies, and direct purchases from private industry vendors. Since data came from a variety of sources, the resolution, area of coverage, file size, and projection varies by dataset. Digital download products are available in a Georeferenced Tagged Image File Format (GeoTIFF).

Images are available at this web address:

https://earthexplorer.usgs.gov/

Setting Up a New Map. Preparing a new map gives you an opportunity to study your map sheet. If you are laminating your map, make these permanent improvements before you seal it up. These additions are for your convenience and ease of use. Any small modification that helps you to find data faster, prevent errors, and assist your personal nav preferences should be added. Be sure to update your declination diagram at a minimum (more on this later).

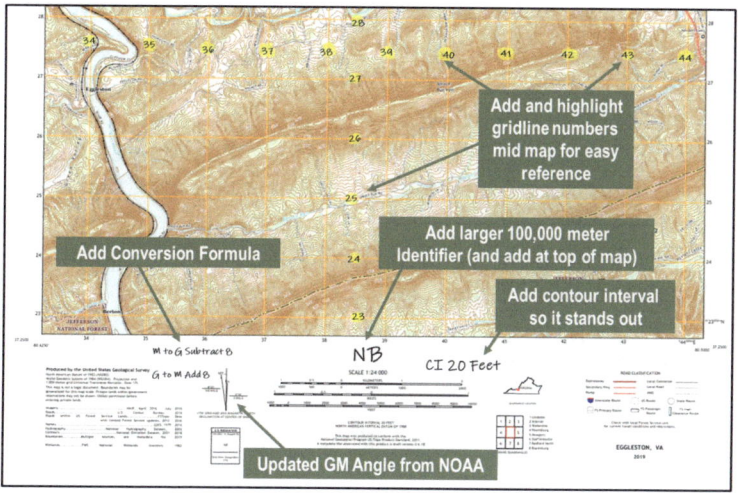

If you trim the margins, it is essential to note any information that may be needed later (such as the map datum and magnetic declination). If you cut the map sheet down any relevant adjacent data must be added back to the modified sheet. You may find yourself in the unenviable position of having to navigate across sheets or even back and forth. Custom cutting and then joining the sheets together is a smart idea in this case. Be sure to have the data that applies to both sheet(s) somewhere on your new cojoined map.

WATERPROOFING

Protecting your map in the field can be done by placing the map in a protective case or by waterproofing liquid or film directly on the map (when setting it up for the first time). Some mapmakers offer the option to print directly on waterproof materials.

Map Cases can be the commercially available type or even heavy-duty freezer bags. Either can work well depending on how you use your map. Some of the larger commercial and military cases can add weight and can be cumbersome when plotting points or adding graphics. Don't go crazy with the newest high speed multicam version of a map case, keep it simple.

For dismounted navigation I always use a heavy mil plastic or freezer bag, less is more sometimes. Mounted land nav is a different story, using larger cases and even map boards are the way to go.

Waterproofing can be spray or brush on chemicals, or a solid film barrier. The advantage of solid film is you can use alcohol markers, but it makes the map difficult to fold. Spray or brush on compounds can sometimes stiffen or yellow over time (although these have gotten better over the years). The downside is you cannot use markers with most types of this waterproofing method.

Do you need a red light readable map? Red light readable means all the map features can be read / seen when using a red lens flashlight. Some maps have contour and feature colors adjusted to appear under red light, some do not. Check the marginal information, if it is not stated be sure to check your map under red light. Use caution using blue or green lenses, there are blue (water) and green (vegetation) inks on maps that can "disappear" under similarly colored lens - which is obviously not a good thing for navigating at night. *Physically check and test your maps with your colored lens choice before you use them.*

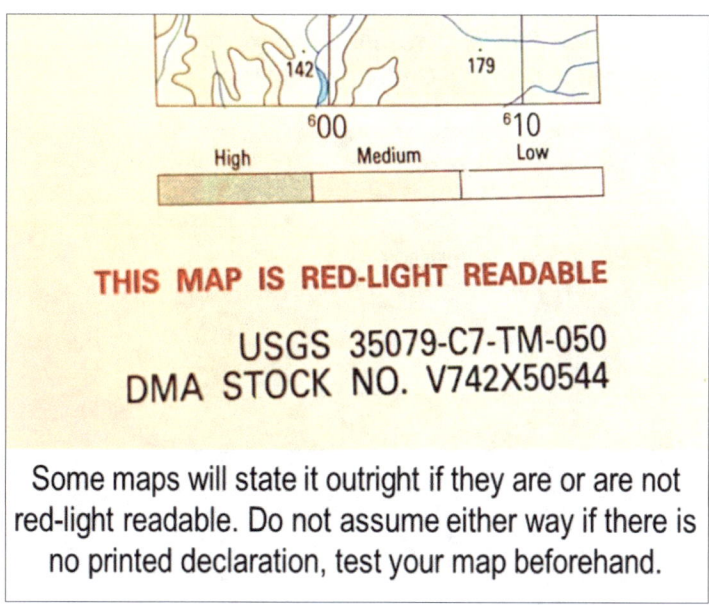

Some maps will state it outright if they are or are not red-light readable. Do not assume either way if there is no printed declaration, test your map beforehand.

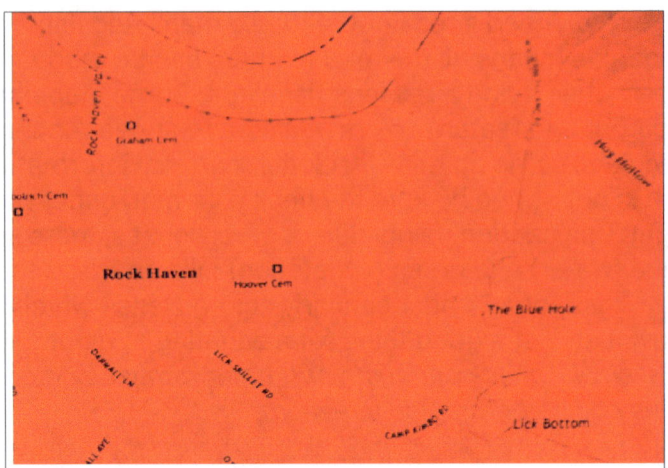

This 2019 USGS Topo is red light readable. Or is it? There are grid lines present in this photo you just can't see them. It is a No-Go for red light use.

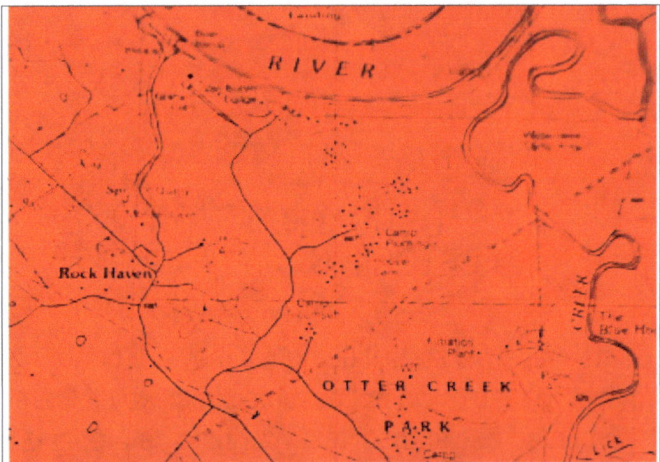

This 2002 Historical topo of the same area is red light readable. Remember the datum may not match, so you must understand all the nuances when you choose your maps. There is not a single optimized solution anymore.

Grid Lines. Current USGS maps have light orange / tan colored grid marks that disappear under a red lens. The contour lines and features will show, but the grid lines do not. MGRS maps and the historical USGS maps do not have this issue. You can get a custom printed map with darker grid lines from some of the online map sites that may or may not solve the issue.

Why could this matter? We all know any light (red lens or even IR) poses a risk of detection at night. However, you may need to read your map with a red lens. Map reading with NVGs can be done, but it is far from optimal. You are better off with NVG and a NV capable GPS unit that you can glance while on the move; but the idea behind learning the manual systems is GPS (our primary) may not be available. And you will *still* need to read your map...GPS or not. I would rather have the red light as an option and not need it. And the newest USGS maps? Yeah, you can't see the grid lines on them with green phosphor NVGs either.

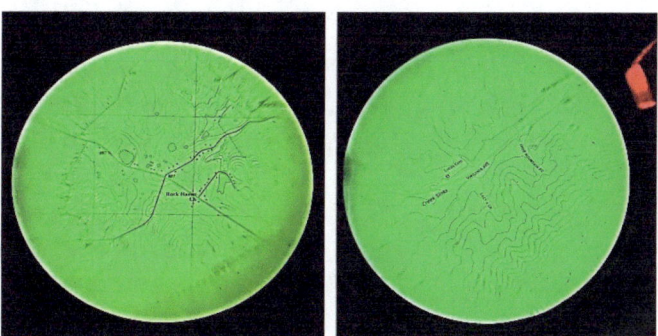

Historical USGS on the left, new map on the right. Both can be read with NV, but the current series topo grid lines do not show under NV. (note the IR light indicator, you may need an IR light source to see details if the lighting conditions are poor; another detection risk)

29

GENERAL MAP KNOWLEDGE

The challenge for a cartographer (a professional map maker) is making a flat representation of a large round object that has surface irregularities. To do so requires a system of reference points and standards to ensure maps are accurate. *This may not seem even mildly interesting to you standing in the cold, rainy wilderness with a map and compass. But you should understand a few of the underlying system mechanics to appreciate the logic behind the map in your hands.*

One of the oldest systematic methods of location is based upon this geographic coordinate system. By drawing a set of east-west rings (latitude or lat) around the globe (parallel to the equator), and a set of north-south rings crossing the equator at right angles (longitude or long) and converging at the poles, a network of reference lines is formed from which any point on the earth's surface can be located. Lines of longitude (meridians) run north-south and measure east-west distances. Lines of latitude run east-west and measure north-south distances. All grid reference systems are based on this approach.

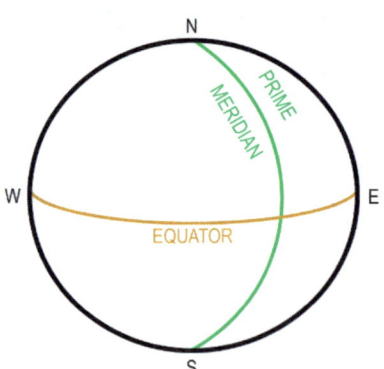

All accurate topographic maps are anchored in a global grid system (Latitude and Longitude).

These lines are duplicated across the globe to create a grid system that allows subdivisions to provide positional coordinates. Mathematically there is an infinite number of refinements, but for navigation anything smaller than 1 meter accuracy is counterproductive. For our purposes if we are within 10 meters (an 8 digit grid) of a navigation objective we will usually have mission success.

Reference System. City streets are named and buildings are numbered, so the only thing needed to navigate (under normal conditions) is the address. Finding a precise location in undeveloped areas or in unfamiliar parts of the world can be problematic. To solve this challenge a uniform and precise system of referencing was developed.

Each "square" or grid on the globe can be subdivided to give us usable location data. *This is not to scale and does not show the true number of grids in the UTM system - it is just for illustration only!*

Each section of grids has a unique identifier and inside every one of those smaller groups there are more identifiers that help refine the location. Think of it like a phone number that starts with a country code, then an area code, then a location code, with final refinement as your specific phone number.

UTM is the **U**niversal **T**ransverse **M**ercator system; it consists of grids superimposed over the earth. It is the standard system for the National Grid and the typical USGS topo maps we use. It is also the basis for the MGRS military map system as well. The UTM is based on lat/long and then further divided into zones and grids.

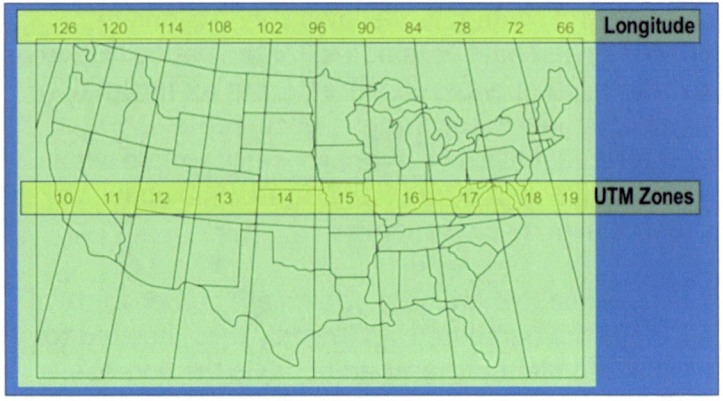

The Universal Transverse Mercator (UTM) grid for CONUS has 10 zones.

A 7.5 minute topo map is barely a dot on the global scale, but with this grid system you can pinpoint exactly where your map sheet belongs on the earth.

The system logically steps down from larger to smaller areas through a series of refinements. Once you understand the "big ideas" of the system these smaller grids and numbers will help make sense of the ones on your map. It is true that you don't need to apply this knowledge to find your way on a map, but context and an understanding of how the system fits together is helpful.

The example on the next page (not to scale) shows how the system flows from the global grid system down to your individual map sheet.

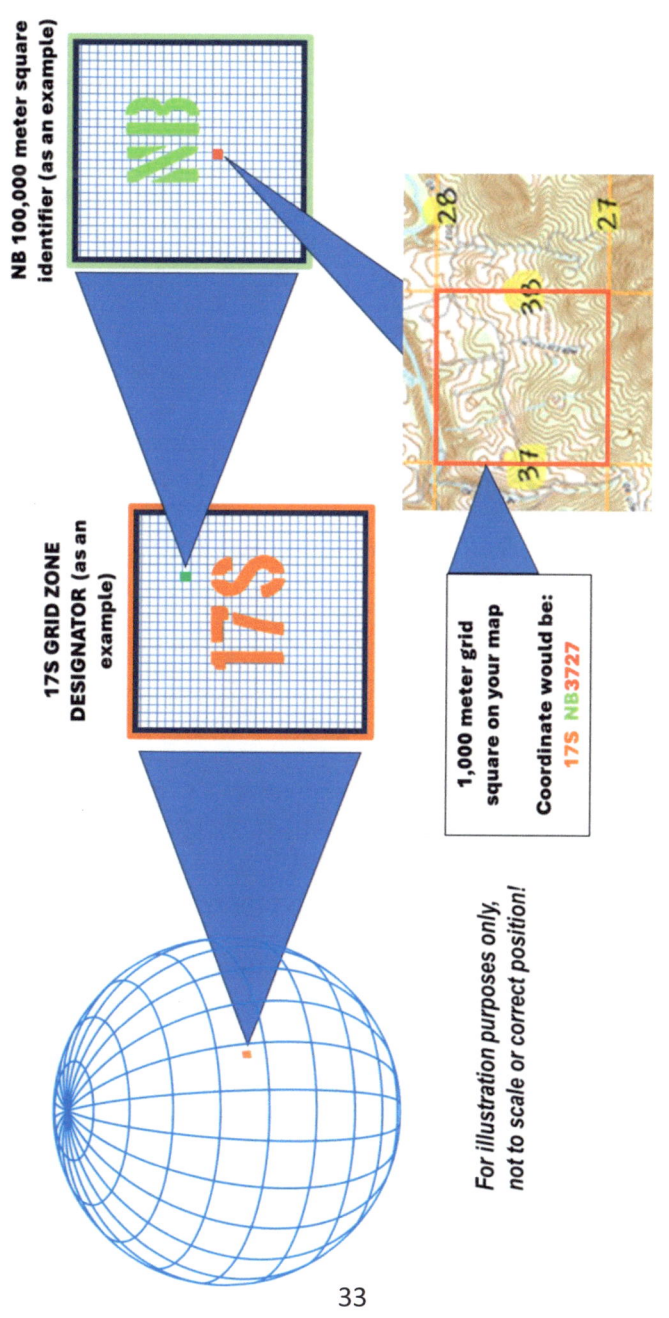

NB 100,000 meter square identifier (as an example)

17S GRID ZONE DESIGNATOR (as an example)

1,000 meter grid square on your map

Coordinate would be:
17S NB3727

For illustration purposes only, not to scale or correct position!

33

UTM vs MGRS. "But my USGS topo map grids won't match, I want MGRS like the Army uses." This is a common misconception, there is no difference in the grid reference system numbering / accuracy printed on the USGS maps. MGRS and UTM grids match up on US maps, so don't worry. The Federal Geographic Data Committee Standard FGDC-STD-011-2001 United States National Grid states:

> *US National Grid (USNG) coordinates shall be identical to the MGRS numbering scheme over all areas of the United States including outlying territories and possessions. USNG basic coordinate values and numbering are identical to UTM coordinate values over all areas of the United States including outlying territories and possessions.*

Grid Squares. The north-south and east-west grid lines on your map intersect at 90°, forming grid squares. No matter what scale your map is the size of one of these grid squares on USGS and MGRS maps is 1,000 meters x 1,000 meters (1 square kilometer or 1k).

Map Datum. A datum is a description of the shape of the earth along with an anchor point for the coordinate system. These are called *map datum* and the specific version used is always annotated on your map. The specific datum is a start point that mapmakers use to build their grid systems. These start points move slightly over time as the earth shifts. Be sure you are using the correct datum in your GPS and datum is consistent across the maps in your team. NAD27, NAD83 and WGS84 are the common map datums in the US.

Datum differences matter. Back to our comparison pic between old and new mapping methods; the two pictures above are the same piece of terrain, the same grid square. Did you notice the grid lines? The historical map on the left uses North American Datum 1927 (NAD27) while the newer one on the right uses North American Datum 1983 (NAD83). Notice how the grid intersection southeast of Hill 2545 has shifted a couple hundred meters south due to using the newer datum.

If you enter the wrong datum in your GPS it can result in significant navigation errors. It will also be a problem if you have hiking partners or teammates using maps with different datum. If you choose to use historical topo maps for your navigation, be sure to account for the grid differences if you are transferring location data to electronics or newer maps.

Depending on where you are in the US the difference between NAD27 and NAD83 can vary from 10 meters all the way up to 200 meters. The difference between NAD83 and WGS84 is less significant at around 1 meter - but you should still account for it.

MAP COLORS

Colors on a standard topographic map are:

1. **Black.** Indicates cultural (man-made) features such as buildings and roads, surveyed spot elevations, and all labels.

2. **Red-Brown.** All relief features, non-surveyed spot elevations, and elevation, such as contour lines on redlight readable maps.

3. **Blue.** Identifies hydrography or water features such as lakes, swamps, rivers, and drainage.

4. **Green.** Identifies vegetation with military significance, such as woods, orchards, and vineyards (these can easily change over time and may be outdated after the map is printed).

5. **Brown.** Identifies all relief features and elevation, such as contours on older edition maps, and cultivated land on red-light readable maps.

6. **Red.** Classifies cultural features, such as populated areas, main roads, and boundaries on older maps.

7. **Other.** Occasionally other colors such as purple on older maps to show new features or pink/red (and others) may be used to show special information such as firing range overlays and impact areas on military maps.

MARGINAL DATA

Marginal data are the things printed outside the terrain portion of the map. This may include map symbols or a legend, the map scale, and other important information such as the declination diagram. Some map prints are very detailed with what they display in the margins, some are sparse. When setting a new map up don't be afraid to add what you need before you waterproof the sheet. If it is value added to you, write it on your map. Do not add operational information such as radio call signs or graphics as permanent map notes.

We need to briefly highlight some items that may be in marginal data. If you want to dive deeper into the details, we recommend reading the Army Land Nav manual FM 3-25.26 and the USGS website for a full explanation of marginal items. Marginal data will vary depending on the type of map you have and even the year / period it was printed; some of these items may not be present on your map at all.

Declination Diagram. This info from this diagram is critical to us as navigators. It shows the relationship in degrees between the easting grid lines on our printed map (grid north) and where our compass points (magnetic north). We will discuss the mechanics of this relationship and the diagram itself in detail inside Chapter 4.

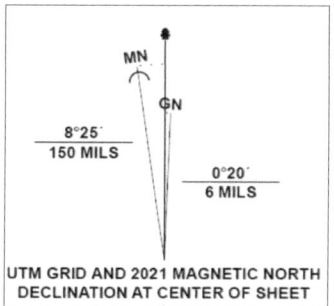

MN

GN

8°25'
150 MILS

0°20'
6 MILS

UTM GRID AND 2021 MAGNETIC NORTH
DECLINATION AT CENTER OF SHEET

Scale. The map scale will be stated in the marginal data, usually at the center bottom of the map. It will also have a bar scale that is useful for measuring distance on the map (it is printed to scale).

Legend. The legend shows the topographic symbols used to depict some of the features on the map. The symbols are not always the same on every map so always refer to the legend. USGS has an extensive legend reference on their website, we have included it as Appendix B in the back of the manual.

Grid Zone Designator and 100,000 meter square identifier. These two pieces of grid info we discussed earlier are always shown in marginal data. As a quick review the grid zone designator is a number and a letter such as "17S" and the 100,000 meter identifier is always two letters such as "NB" or November Bravo.

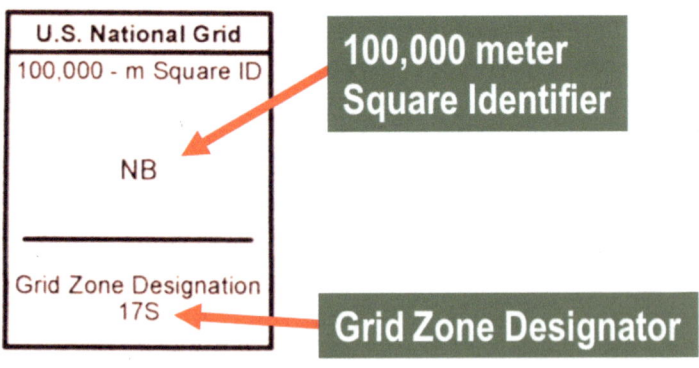

Adjoining Sheets Diagram (MGRS) or Adjoining Quadrant Name (USGS). A grid that shows the map sheet numbers or names surrounding your current map sheet. These are helpful when determining which maps you need to acquire prior to a mission.

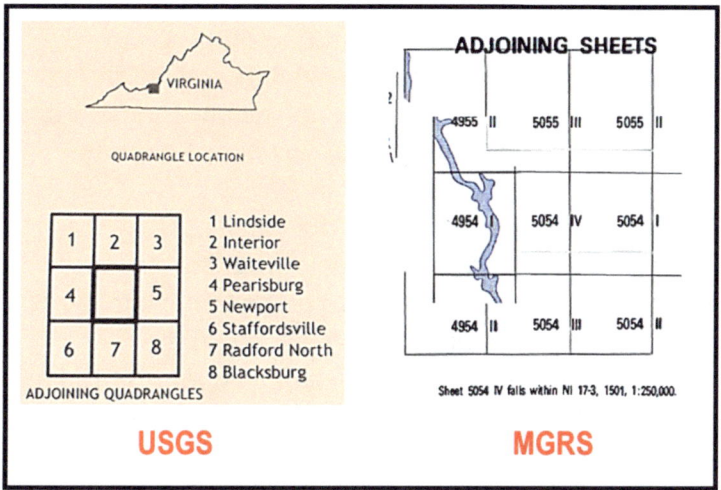

The Adjoining Quadrangles diagram on USGS topo will show the names of the map sheets next to you. This is convenient for ordering or researching terrain around your area. The MGRS maps have an "adjoining sheets" diagram that carries the same type of data.

Contour Interval. The elevation gain between two contour lines is always stated on a map, usually near the scale. This will vary depending on the terrain (maps with a lot of steep terrain will have larger intervals to keep the number of contour lines manageable). It will be stated as Meters or Feet. Larger contour intervals mean less detail on your map.

COMPASS

A compass is a precision instrument that points to magnetic north. Of the dozen or so compass types there are a few that will meet our set of prepper/readiness or small unit / group leader requirements. The lensatic, baseplate, and electronic compass are your go-to options. The environment and conditions we face requires us to choose a durable, reliable, and accurate compass. We cannot afford to have low quality nav items in our kit, especially when it comes to compasses. You know what that means when it comes to cost, you are going to have to lay out some money for a good compass. Avoid the temptation of the $6 dollar baseplate compass (as of publication date there are several of these $5 and $6 models through the popular online resellers). They may resemble the good ones in appearance, but they are junk. Part two of this story is you need a backup compass as well…so it is going to hurt twice as much to buy good quality.

Lensatic compasses. The military grade lensatic should be your primary compass for serious navigation work. A lensatic is a card compass which means the magnetic north seeking needle and the scale (the numbers in degrees) both rotate together vs having an outer ring that rotates independently. There is a magnifying lens on the rear sight used to see the card dial details to help take an accurate reading. The presence of this magnifying lens is what makes it a "lensatic" compass.

There are two illumination options for military lensatic compasses. Tritium and phosphorescent illumination each have their advantages, but my preference is always tritium. It requires no "charging" with light to glow. The downside is the tritium will fade as the radioactive material decays over time, so have a plan to replace a compass every 8-10 years.

Lensatic Compass Components

General Inspection. Compasses are precision instruments and should be cared for accordingly. Inspect your compass before each mission or outing. Some recommended serviceability checks are outlined on the next page.

Lensatic Compass Visual Inspection Checklist

Cover - Make sure the cover is securely attached and is not damaged or cracked. The cover protects the compass from damage and serves as a sighting device.

Sight - Check that the sighting wire and the sighting slot on the lens assembly are working properly and is not damaged.

Lens - The lens should be clean, free of cracks, scratches, or other damage.

Bezel - The rotating bezel should rotate smoothly and easily. The markings on the bezel should be clearly visible and not worn away.

Needle / Dial - The magnetic needle should be free of rust or corrosion and should move easily in response to magnetic fields. The dial should be clearly marked and easy to read. It must rotate smoothly and easily over the dial without any slipping or jumping.

Housing - The housing of the compass should be straight and damage that could affect the accuracy of the compass readings.

Baseplate compasses. These are designed for hiking, orienteering, and other outdoor activities. They feature a flat baseplate with a rotating bezel, a magnetic needle, and a ruler for measuring distances on maps. The baseplate is usually a fast-settling compass that is outstanding for orienteering and terrain association. They are not the optimal tool for dead reckoning or obtaining a precise azimuth, the lensatic is best when it comes to those tasks. Quality baseplates are usually liquid filled; when inspecting ensure there are no bubbles present in the sight glass as these can disrupt the needle float. Even a small bubble will throw an azimuth off as the needle settles. A Sighting Compass is a hybridized type of baseplate that adds a mirror and a sight of some type. Silva Ranger and Suunto MC2 are examples of these.

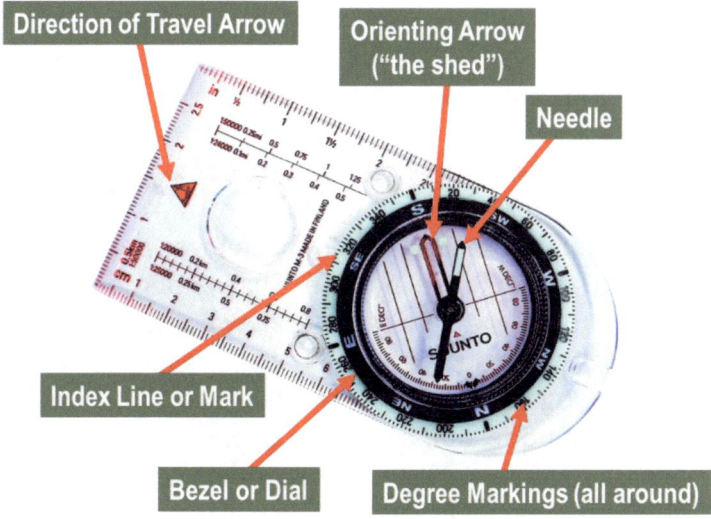

Global needle. Some baseplate compasses have a global needle. Compass needles are balanced so they will not drag on the housing as it rotates. A compass balanced for the northern hemisphere will drag or stick in the southern magnetic zone. Global compasses have a different pivot style for the needle assembly, along with a taller compass capsule that allows the global needle to rotate freely even when it isn't completely level. So what? You aren't headed south of the equator anytime soon. A global needle can be a good feature to have when orienteering or moving quickly with a baseplate compass. Not having to level the compass precisely makes it easier to take an accurate reading while you're still moving (within reason).

Declination. Some baseplates have an option to set the magnetic declination on the compass itself. My advice is do not use this feature, always do the math for grid conversion (more on this later). The convenience of having it preset is not worth the potential confusion it can cause by having two sets of procedures - especially when you are tired.

Never buy an off-brand baseplate compass to save a few bucks. Silva, Suunto, and Brunton are all among trusted brands, but even within these there are models known to have issues. There are dozens of models with varying features that may or may not work for you. Get some field time with your choice and it will become evident if you like a particular model. We recommend choosing a model that has the phosphorescent markings for use at night. Some models (even the well-built ones) may not have these.

Azimuth Check. This technique is for both lensatic and baseplate compasses. Use a horizontal flat or vertical square surface free of metal (including nails). Measure an azimuth to the nearest degree using a calibrated electronic compass (a calibrated smart phone can work great for this). Record the azimuth and check your compasses against the baseline. A one- or two-degree difference should be the max. Do this with new compasses, if they are out of the manufacturer's stated tolerance get a replacement. Two or more baselines with a minimum of 90 degrees difference should be used.

Check each of your compasses on at least two azimuths. This can be done on a flat horizontal surface as well. The area must be outdoors, away from metal or other compasses. Do not use a concrete surface like a driveway or patio since the metal rebar in the concrete can interfere with the process.

45

GPS compasses. These use GPS technology to determine direction. They are useful in areas where traditional compasses are not effective, such as in the presence of strong magnetic fields and large metallic objects. This means you can rely on them when mounted (when you are on a vehicle). They require a GPS signal so if the GPS system is disrupted your compass will not work. True GPS compasses require the unit to be moving to function properly, usually faster than 2 mph.

Electronic Compass. An electronic compass functions much like a traditional compass, and just like a traditional it can be negatively influenced by magnetic or metal objects. It can be used when stationary or at slow speeds. The device must be held level to give an accurate reading. An electronic compass may require calibration from time to time. Electronic compasses can also be in GPS units so just because you are using a GPS unit does not indicate it is a GPS based compass.

3-Axis Electronic Compass. – This type of electronic compass can be used when stationary or moving. This compass is tilt compensated, meaning it does not have to be held level to provide an accurate reading. As with the standard electronic compass it may require calibration. They are magnetic, so keep them away from metal when using them as you would any other standard compass.

A backup compass can be another lensatic or a smaller, lighter baseplate compass. My personal preference is having a baseplate stashed in a shoulder or cargo pocket as an alternate. If I am using terrain association only or just doing orienteering, I will use the baseplate compass as my primary (more on this later when we discuss staying on the route).

A small watchband or wrist compass is a great contingency item, but these should not be second in line if your primary lensatic becomes non mission capable. The PACE (Primary, Alternate, Contingency, Emergency) system applies to navigation tools as well, especially for compasses.

Military Surplus NVG Compass. These are garbage, if you have NVGs don't waste your time or money on them. They were a good idea poorly executed.

The only place these NV "compasses" should go. Don't waste your money.

47

PROTRACTOR OR SCALE

Protractor. There are several types of protractors—full circle, half circle, square, and rectangular. All of them divide the circle into units of angular measure, and each has a scale around the outer edge and an index mark. The index mark is the center of the protractor circle from which all directions are measured. If you choose a circular protractor, be sure to have a straight edge in your kit as well. Full size square protractors will usually have Mils on the outside and degrees on the inner scale. Some navigators will trim off the mils which is more convenient. If you chose to cut the mil scale off be precise when trimming so you maintain the straight edge.

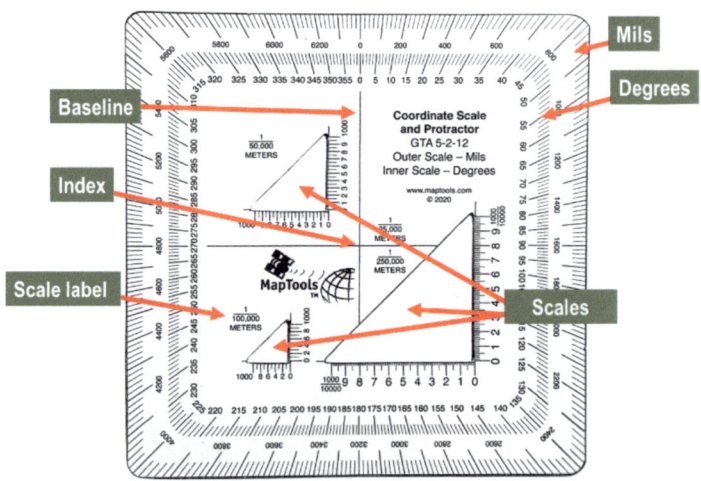

Coordinate scale and protractor. There are many aftermarket brands like this MapTools version that work well. The army issued protractor is actually one of the worst options (least durable and poorly cut). You must have a protractor that matches your map scale, the military coordinate scale does not work on a USGS 1:24000 map.

Protractors with multiple scales are great, just make sure you are using the correct set of numbers that matches your map. It is easy to make a mistake and use the wrong scale when you are cold, wet, and out of your mind tired. Circular protractors are excellent for annotating azimuths on a map, just ensure you have an additional straight edge to use with them.

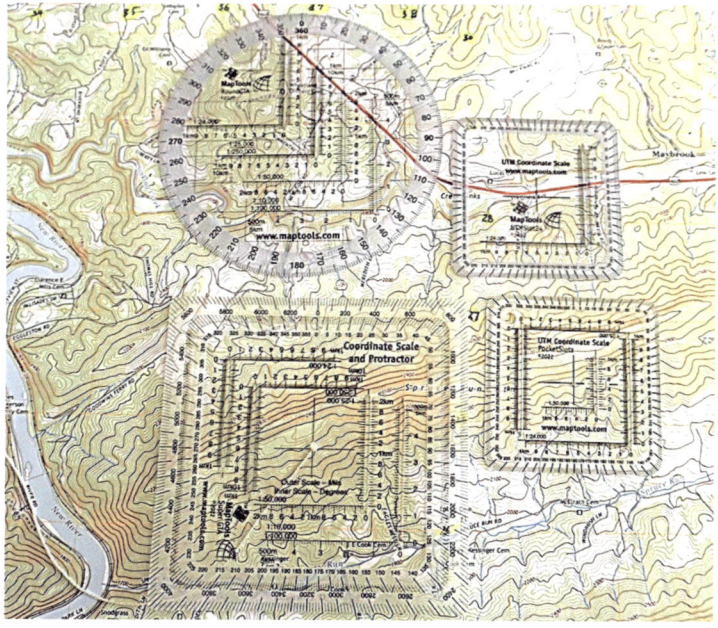

Protractors or scales are required for precise navigation on a topo map. Have a good quality, properly cut protractor that matches your map scale and personal preference.

A quick note on 1:50,000 Coordinate Scale. The 1:50,000 scale is subtended in 50 meter segments. Using interpolation, you must mentally divide each 50-meter segment into tenths. For example, a point that lies after a whole number but before a short tick mark is identified as 10, 20, 30, or 40 meters and any point that lies after the short tick mark but before the next whole marked number is identified as 60, 70, 80, or 90 meters (the associated 8 digit grid with these only uses the first number, more on this later). The 1:50,000 scale below is typically used on MGRS or military maps.

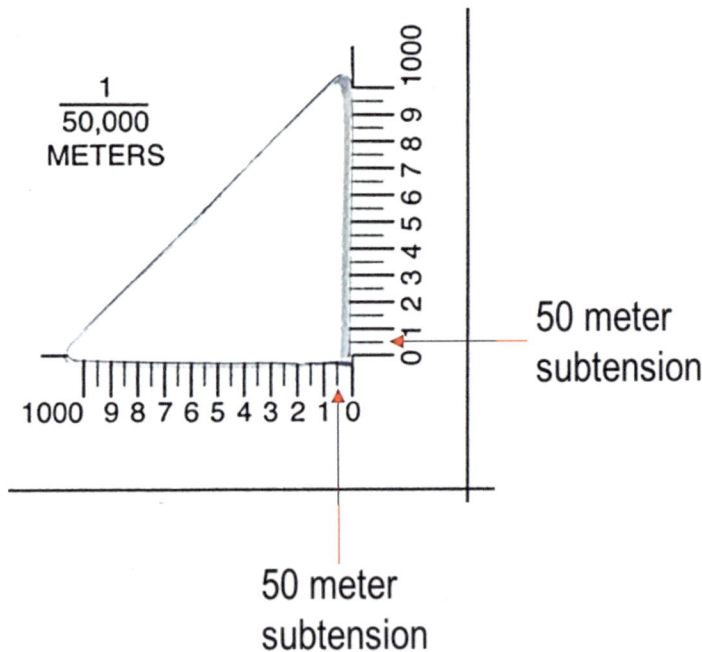

GPS

The Global Positioning System (GPS) refers to a specific US based navigation satellite system. However, we have all come to refer to any satellite nav device or system as "GPS" so we will stick with that widely accepted inaccuracy for this manual.

GPS units are small radio receivers that pick up low power signals from satellites. They require signals from at least four satellites to determine a location. The units we are focused on are handheld or wrist worn units with an on-board power source (either replaceable AA/AAA or permanent rechargeable).

The GPS is and should be your primary means of land navigation; you must have a basic (at a minimum) GPS unit in your kit. A smart phone app does not equate to a handheld GPS unit. Your device can be a backup GPS in your PACE plan, just be sure to account for the negative consequences of relying on a mobile device in the backcountry...and the tracking vulnerabilities of mobile phones.

Modern GPS units are easy to use, relatively inexpensive, power efficient, and reliable. Handheld GPS units have followed personal computing trends by decreasing in price and increasing in performance.

Before inflation grabbed hold in 2022 you could scoop up a base model Garmin eTrex for around $89 bucks, but even at $150 inflationary pricing they are still well worth it.

GPS VULNERABILITIES AND RISKS

The signals from nav satellites are relatively weak and can be disrupted by terrain or overhead vegetation. Jamming, spoofing, encryption, denial of service hacking and satellite killers (loss of system) are all potential GPS disrupters. These risks reinforce our need for map and compass proficiency. It doesn't take a World War III satellite shoot down or an EMP to cause issues, something as simple as a lost or broken unit in the woods can take GPS capability away from us. A misconception is a GPS unit can be tracked; it is a one-way system and cannot be located (but other components may be). GPS only transmits, it requires no signal back from your actual device to work.

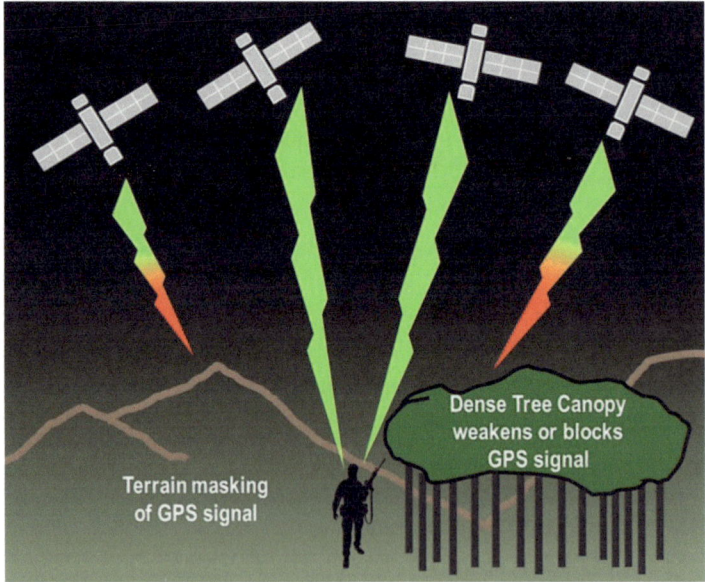

Dense Tree Canopy weakens or blocks GPS signal

Terrain masking of GPS signal

Your GPS must have the signal from four (4) satellites for accurate position data. Terrain (including tall buildings / urban canyons) and overhead vegetation can disrupt or further weaken an already weak signal.

Jamming and Spoofing. The weak signals from nav satellites are vulnerable to jamming. There are other codes and frequencies for military GPS use, the new M Code being among them that are less vulnerable to spoofing or jamming. Any bad actor can easily jam the three civilian authorized signals locally with relatively inexpensive technology. Spoofing is when a threat hijacks the carrier signals and retransmits false versions to GPS receivers. This is unfortunately becoming more frequent. There is a website that tracks these incidents worldwide, it is worth looking at to appreciate the system risk.

https://gpsjam.org

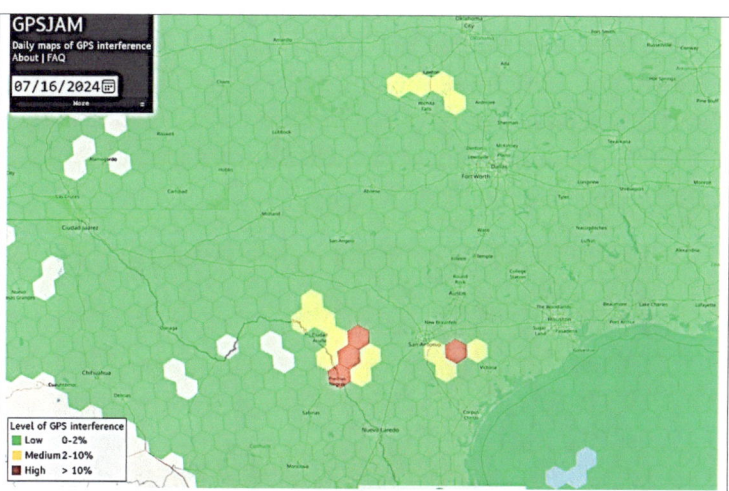

GPS jamming reports from 16 July 2024. The southern border is lit up a bit, if you look at the European reports for jamming and spoofing it is even more prevalent. (source gpsjam.org)

Several countries have put navigation satellites systems in orbit (each system is referred to as a constellation). GLONASS is Russia's system, BeiDou belongs to China, QZSS is the Japanese regional system, and Galileo was developed by the European Union. Some GPS units have multi band capability which means they can use these other constellations. GNSS (Global Navigation Satellite System) is a term that refers to this international multi-constellation system.

The downside of using GNSS is it will consume more power than just using the GPS constellation. The upside is your fix time (first good location) will be reduced and your location data can be more accurate since the more satellites your GPS tracks the better. Multiband units have the option to turn this feature on and off to save power. GNSS can also mitigate the risk of GPS spoofing since it is difficult for a bad actor to spoof multiple systems at once.

HANDHELD GPS FEATURES

ABC Sensor. Altimeter, Barometer, Compass. The compass can be either GPS, electronic or both. Some wrist worn devices that have GPS units will include these features and some non-GPS watches will have ABC. These can be used to augment your handheld GPS or be part of your navigation PACE plan.

Maps. Some units have maps as a feature. These can be helpful and convenient, but their utility can be limited if you are expecting too much out of them on a small screen. Some units have maps that reside on the unit, some rely on connectivity to display the map data.

Power. Modern GPS units have become very power efficient, some can run hundreds of hours on a single set of batteries or charge. Units are available with replaceable batteries (usually AA or AAA) or they can a have a permanent rechargeable battery. Account for these options and differences in your power management plan and battery commonality.

Direct Sunlight Readability. This is a big one. There are popular units that struggle with visibility in bright sunlight. Ensure you read and consider customer reviews of the screen contrast when you select a GPS unit. Color screens can appear more washed out than high contrast monochrome displays, but some full color maps do surprisingly well in sunlight. We recommend you research and get your hands on the specific model before you commit to a unit with a color display.

Touch screens. I'm not a fan of touch screens on a handheld GPS unit for field use. There are high-quality touch screen handhelds on the market, just consider the environment and your anticipated use case if you are considering one. Remember gloves, rain, cold, snow and mud all get a vote.

Multi Band. This option allows a unit to access some or all the international GNSS constellations in addition to the GPS constellation.

Satellite Comms / SOS. Some GPS units have satellite communication as a feature. Having the ability to message a loved one your status on a trip or use the SOS function to call for SAR (Search and Rescue) in an emergency can be of great value especially for long trips in the backcountry. METT-TC applies, this feature may not be applicable in some crisis conditions but for the peacetime hiker with a busted leg these can be literal lifesavers. The systems require a subscription, fortunately the good ones allow you to cycle the monthly subscription on and off as needed. Some of these can be used to navigate, but the small units are not optimized to be a primary GPS. It is best to use these mini types as part of an overall system and have a separate primary GPS.

Waterproof or Water Resistant? This is a good time to talk about water resistance ratings of electronics. *Waterproof* means the item can be fully submerged. Depths and submersion time underwater vary by product type. *Water-resistant* means the item can take splashes and light rain, but not full submersion.

IP ratings indicate specific protection for both solids (dust) and water. Verify your electronic ratings before exposing them to the elements.

Ratings. IP stands for International Protection (or Ingress Protection). Ratings indicate how much protection an item has against solids (the first number), and liquids (second number). The higher the number, the more resistant it is to dirt or water. The IP rating is always "IP" followed by two letters or numbers. An "X" in either space means that item isn't rated for that type (solid or liquid) of protection. For example, "IPX2" means that the item isn't rated for dust or solids but has a water resistance rating of 2. The next page has all the rating numbers and what they mean for your GPS, rangefinder, or electronic compass.

SOLID PROTECTION	
X	No data
0	No protection against solids
1	Solids greater than 50mm can't enter
2	Solids greater than 12.5mm can't enter
3	Solids greater than 2.5mm can't enter
4	Solids greater than 1mm can't enter
5	Keeps most dirt and dust out
6	100% dust-proof

LIQUID PROTECTION	
X	No data
0	No protection against water
1	Vertically dripping water
2	Water dripping at several angles
3	Spraying water
4	Splashing water from all directions
5	Water projected from a 6.3 mm nozzle at low pressure
6	Water projected from a 12.5 mm nozzle at high pressure
7	Immersion into up to 1 meter of water for 30 minutes
8	Immersion more than 1 meter for a time defined by the manufacturer

NAVIGATION KIT

Components are different for each of us, but there are some basic items we should always have in our nav kit. The field items always apply so we won't add those to this list (IFAK, lights, signaling, water etc). Appendix A of this manual has a comprehensive navigation gear checklist you can modify and use to manage your setup.

In addition to the primary items we discussed, there are additional items you will need. Tailor and modify these to fit your requirements:

Notebook. Always use a water-resistant notebook.

Pencil. Mechanical .7 to .9 mm works best. Also have a backup, either a mechanical or number 2.

Superfine alcohol marker. If you use a laminated map a superfine marker is good to have for marking points or nav objectives.

Alcohol Eraser. If you use markers this is also a requirement.

Pace Beads. These are the best method to keep track of your pace during a navigation leg.

Rangefinder. A laser rangefinder is useful to determine the distance to features on the ground.

The basics of a "land nav kit". Calling it a kit is a bit inaccurate since these are not standalone items and should always be part of a wilderness or tactical loadout.

CHAPTER 2
KNOW WHERE YOU ARE

You must be able to read a map (understand and interpret the symbols and features) to know where you are. This chapter covers the foundational navigation skills of determining locations on a map, measuring distances and azimuths, and identifying symbols on a map.

GRIDS

Plotting your location on a map is the only way to build the complete picture of your relationship to the terrain. Reporting positions to another team, a hunting partner, or a search and rescue team is critical. We will use the standardized grid system introduced in Chapter 1; it provides us with shared language and a common reference to send accurate information to others.

Knowing where you are (position fixing) and communicating that data is crucial to navigation. It is essential for accurate reporting of various dangers or assisting with search and rescue efforts. Few factors contribute as much to your survivability and successful mission accomplishment as always knowing where you are.

Locate a Point Using Grid Coordinates. Based on the military principle for reading maps (RIGHT and UP), locations on the USGS or MGRS maps can be determined by grid coordinates. We always read map coordinates EAST then NORTH (RIGHT then UP).

To locate or plot a point on a topographic map you will use a coordinate scale, commonly called a protractor. These coordinate scales can be the military Graphic Training Aid (GTA) type or a commercial version.

Ensure you have the scale that matches your map. Scales can vary for maps; the one consistency will be the measurements of angles. Angular measurements for land nav will be in degrees (most common) or milliradians (mils).

Military (MGRS system) topo maps are usually 1:50,000 or 1:25,0000 scale with some 1:100,000. The USGS (US Geological Survey) 1:24,000 scale maps are most likely what we will have access to and use as normal folk. MGRS maps may be available during a crisis, but the techniques for the USGS map will apply to these as well. For USGS 1:24 maps you will need to acquire a 1:24,000 protractor; **the standard military GTA (Graphic Training Aid) coordinate scale will not work for USGS 1:24,000 maps.**

The number of digits in a complete grid represents the level of precision to which a point has been located; the more digits the more precise the measurement.

A four-digit grid is accurate to 1,000 meters (a full grid square), a six digit grid will get us within 100 meters of a location, an 8 digit grid within 10 meters, and a 10 digit within 1 meter (10 digit grids are GPS only, paper map and protractor will realistically get you to 8 digit resolution). The more errors we can remove (math errors, pencil or marker thickness, protractor placement on the map etc) the more precision when marking and measuring a location.

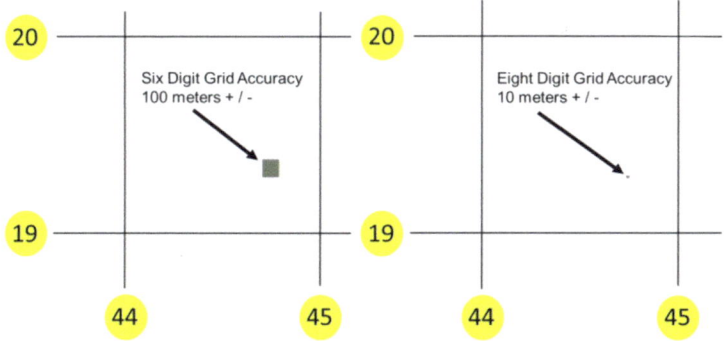

Use a Coordinate Scale (1:24,000 scale example)

When using a coordinate scale for plotting grid coordinates, ensure that the appropriate scale is being used for your map and that the coordinate scale is right side up. On the coordinate scale, there are two axes on the triangle or backwards "L" shaped scale: the horizontal (Easting or first reading – read RIGHT) and vertical (Northing or second reading – read UP). These sides are 1,000 meters in length. The point at which the north/south and east/west sides of the scale meet (the bottom right corner of the scale triangle) is the zero-zero point. Each 1,000 meter side is divided into ten equal 100-meter segments by a long tick mark and number. On a 1:24,000 scale each 100-meter segment is subdivided into 10-meter segments by a short tick mark with a midpoint (or 50 meter) line between each numbered line. To ensure the scale is correctly aligned, place it with the zero-zero point at the lower left corner of the grid square.

Keep the horizontal line of the scale directly on top of the east-west grid line and slide it to the right until the vertical line of the scale touches the desired point on the map. Keep the protractor "squared up" by ensuring the horizontal line of the scale is aligned with the east-west grid line and the vertical line of the scale is parallel with the north-south grid line. To locate the point to the nearest 10 meters, measure the hundredths of a grid square RIGHT and UP from the grid lines to the point. The point in the figures is in the 4520 grid square (remember complete grids are always an even number). The coordinates to the nearest ten meters are 45352016. If a six-digit grid was given the grid would be 454202 and would be accurate to 100 meters.

How reading or plotting a grid coordinate looks with just the scale and grids and then on the 1:24,0000 map. Keeping the scale squared up is critical:

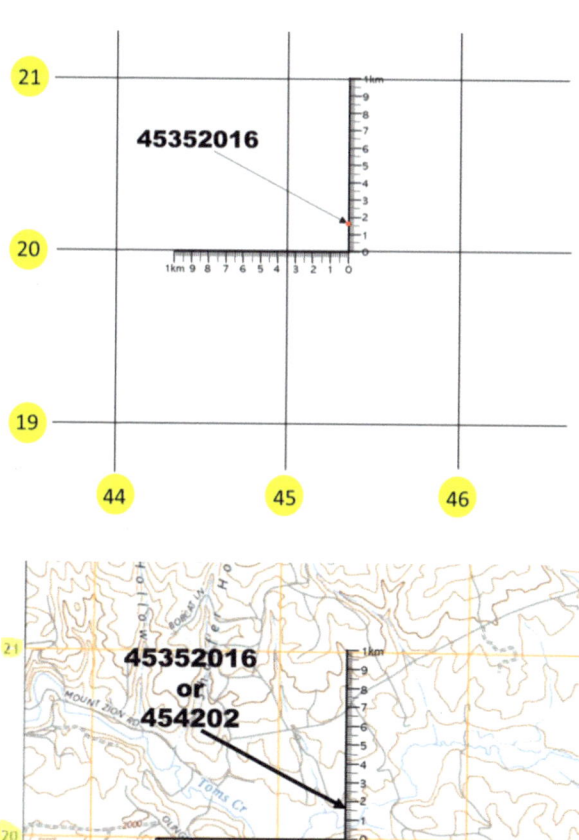

Example using a 1:50,000 map and the associated scale on the protractor:

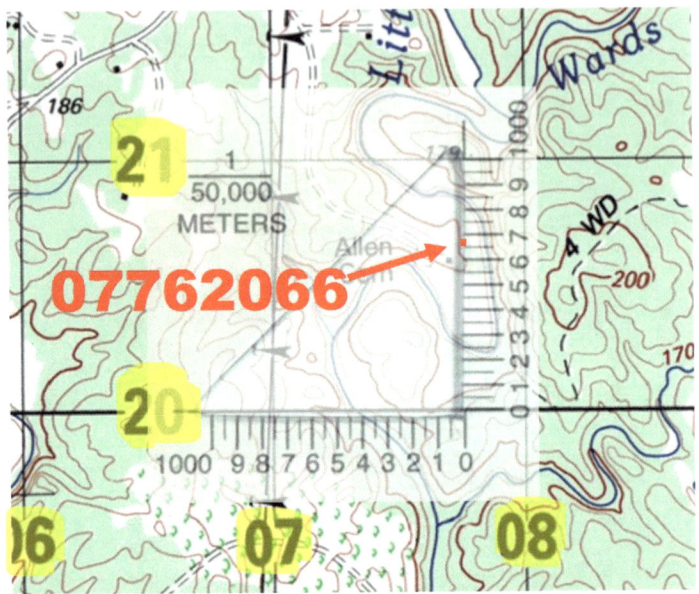

1:50 scale using interpolation to get an 8 digit grid. The easting is between the 750 and 800 meter hash marks; closer to the 750 so we will estimate 760. The northing is estimated to be just above the 50 meter subtension between the 6 and 7 hundred meter lines on the vertical scale so we can estimate it at 660 meters. (since it is an 8 digit grid it is written as 6 and not 60)

Recording and Reporting Grid Coordinates.
Coordinates are written as one continuous number without spaces, parentheses, dashes, or decimal points; *grid coordinates always contain an even number of digits.* Know where to make the split between the RIGHT (easting) and UP (northing) readings for written coordinates. Since there are thousands of duplicate coordinate numbers across the full series of map sheets it is important that you include the 100,000-meter square identifier letters (described below). We typically determine grid coordinates to the nearest 100 meters (six digit grid) for expedient reporting and 8 digit for more precise locations.

100,000 Meter Square Identifier. Every map sheet is part of a larger grid system (as discussed in Chapter 1). The identifier is part of the marginal information on the map, it is a two-letter code. To prevent confusion when doing work across map sheet boundaries we use the 100,000 Meter Square Identifier at the beginning of the grid coordinate. As an example, we would use "NB" or "November Bravo" when referencing grids on a map sheet with that identifier - so the six digit grid from the prior example is NB 454202. An identifier changes across each double zero "00" easting and northing, so be aware you may have two (or even four) 100,000 Meter square identifiers on one map sheet. If working internally and using a single map sheet we will oftentimes eliminate the identifier when reporting grids.

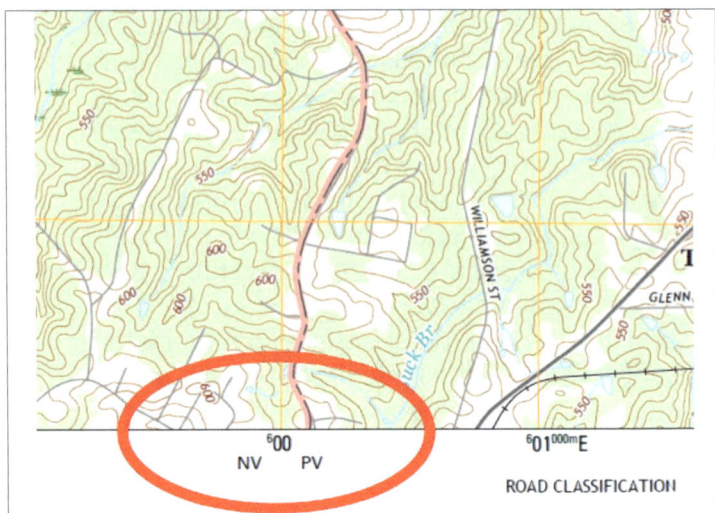

An example of a single map sheet that has more than one 100,000 meter square identifier. As you cross the 00 easting on this map sheet it changes from NV to PV. The same will be true on a map sheet that has a 00 northing.

The 100,000 meter square identifier is mistakenly (even by some professionals) referred to as the 'Grid Zone Designator". The Grid Zone Designator is a set of numbers and a letter such as "17S" or "15A" to further identify where a series of map sheets falls into the larger grid scheme. It is one level above the 100,000 meter square identifier and is seldom used for our purposes other than locating and acquiring the proper map sheets for an area.

TERRAIN FEATURES

Understanding the grid overlay on the map is just the beginning. The contour lines drawn on a map convey a story or snapshot of what you will encounter on the terrain. A topographic map is an accurate representation of the ground, but we must know how to "read" or interpret the map to use it properly. Learning to read a map is like learning a new language. With practice it will become second nature to you and the map features will make sense as your brain pulls it all together. All the lines and shapes arranged on a map will speak to you; the meaningless squiggles and circles suddenly become recognizable terrain features.

All terrain features are derived from a complex land mass known as a mountain or ridgeline (see figure below). The term ridgeline is not interchangeable with the term ridge. A ridgeline is a line of high ground, with changes in elevation along its top and low ground on all sides from which a total of 10 natural or man-made terrain features are classified.

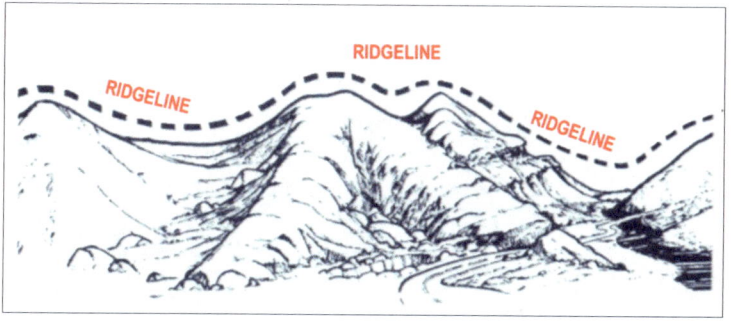

FIVE MAJOR TERRAIN FEATURES

Hill. A hill is an area of high ground. From a hilltop, the ground slopes down in all directions. A hill is shown on a map by contour lines forming concentric circles. The inside of the smallest closed circle is the hilltop.

Saddle. A saddle is a dip or low point between two areas of higher ground. A saddle is not necessarily the lower ground between two hilltops; it may be simply a dip or break along a level ridge crest that is considered as a pass. If you are in a saddle, there is high ground in two opposite directions and lower ground in the other two directions. A saddle is normally represented as an hourglass.

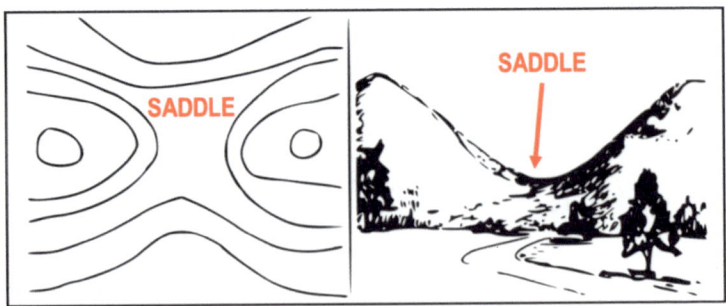

Valley. A valley is a stretched-out groove in the land, usually formed by streams or rivers. A valley begins with high ground on three sides, and usually has a course of running water through it (solid blue line for persistent water, dashed/dotted line for an intermittent (seasonal or wet weather) stream. If standing in a valley, three directions offer high ground, while the fourth direction offers low ground. Depending on its size and where a person is standing, it may not be obvious that there is high ground in the third direction, but water flows from higher to lower ground. Contour lines forming a valley are either U-shaped or V-shaped. To determine the direction water is flowing, look at the contour lines. The closed end of the contour line (U or V) always points upstream or toward high ground.

Ridge. A ridge is a sloping line of high ground. Standing on the centerline of a ridge, you will have low ground in three directions and high ground in one direction with varying degrees of slope. If you cross a ridge at right angles, you will climb steeply to the crest and then descend steeply to the base. When you move along the path of the ridge there may be either an almost unnoticeable slope or an obvious incline. Contour lines forming a ridge tend to be U-shaped or V-shaped. The closed end of the contour line points away from high ground.

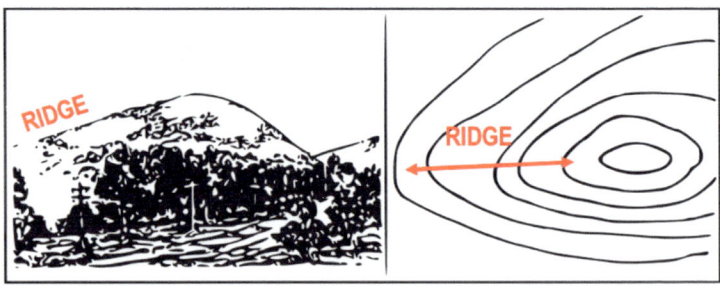

Depression. A low point in the ground or a sinkhole surrounded by higher ground in all directions. Only depressions that are equal to or greater than the contour interval will be shown. They are represented by closed contour lines that have tick marks pointing toward low ground.

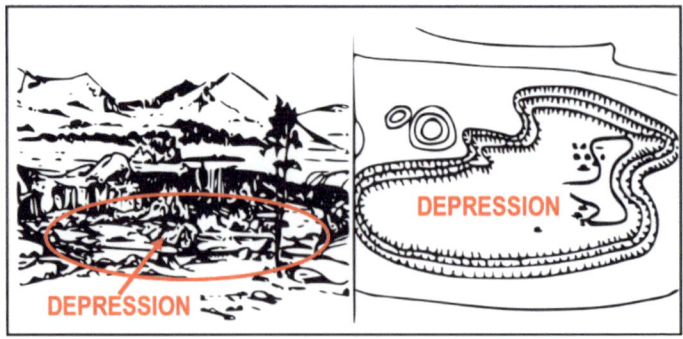

THREE MINOR TERRAIN FEATURES

Draw. A draw has a less developed stream course than a valley. In a draw there is essentially no level ground and it has limited maneuver space within its confines. If you are standing in a draw, the ground slopes upward in three directions and downward in the other direction. A draw could be considered as the initial formation of a valley. The contour lines depicting a draw are U-shaped or V-shaped, pointing toward high ground. In woodland areas draws will tend to have thick vegetation and rugged ground that is difficult to walk through. Crossing multiple draws is physically taxing, and cutting across them at night gets old in a hurry. Depending on the environment they can provide good covered and concealed routes, but they have the tactical disadvantage of being surrounded by high ground on three sides.

Spur. A spur is a short, continuous sloping line of higher ground, normally jutting out from the side of a ridge. A spur is formed over geologic time by two parallel streams (water may or may not be present today), which cut draws down the side of a ridge. The ground sloped down in three directions and up in one direction. Contour lines on a map depict a spur with the U or V pointing downhill / away from high ground. In wooded terrain the spurs are usually not as thick with vegetation as the draws are.

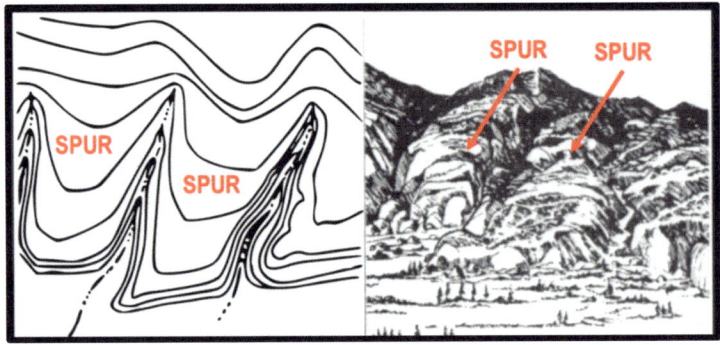

Cliff. A cliff is a vertical or near vertical feature; it is an abrupt change of the land. When a slope is so steep that the contour lines converge into one "carrying" contour of contours, this last contour line has tick marks pointing toward low ground. Cliffs are also shown by contour lines very close together and, in some instances, touching each other. You and everyone in your team needs to be aware of when maneuvering near a cliff, especially at night.

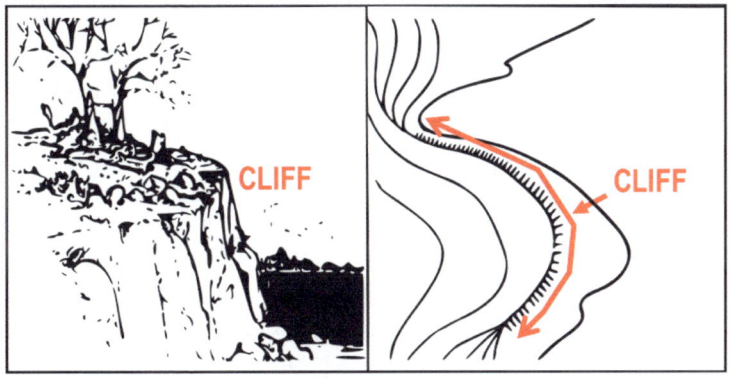

CLIFF (Defined)

CLIFF (Indicated by Contour)

TWO SUPPLEMENTARY TERRAIN FEATURES

Cut. A cut is a man-made feature resulting from cutting through raised ground, usually to form a level bed for a road or railroad track. Cuts are shown on a map when they are at least 10 feet high, and they are drawn with a contour line along the cut line. This contour line extends the length of the cut and has tick marks that extend from the cut line to the roadbed, if the map scale permits this level of detail. Cuts and fills are often adjacent to each other, but not always. Some people mistakenly think it is one feature "cut and fill".

Fill. A fill is a man-made feature resulting from filling a low area, usually to form a level bed for a road or railroad track. Fills are shown on a map when they are at least ten (10) feet high, and they are drawn with a contour line along the fill line. This contour line extends the length of the filled area and has tick marks that point toward lower ground. If the map scale permits, the length of the fill tick marks are drawn to scale and extend from the base line of the fill symbol.

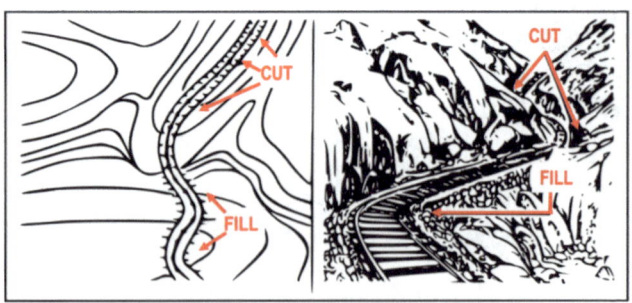

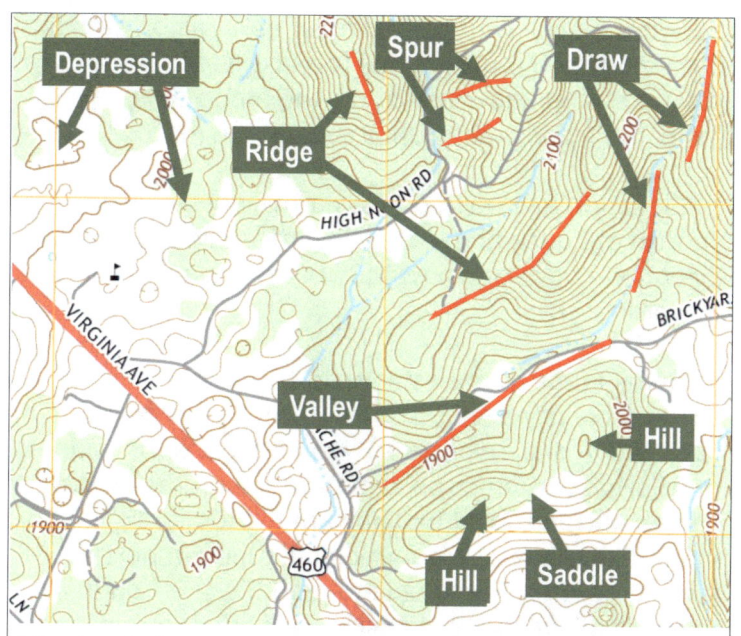

A few terrain features and how they appear on a typical topo map in the Eastern woodlands. (the red lines are added to emphasize the general trace of the features)

ELEVATION AND RELIEF

The elevation of points on the ground and the relief of an area affects the movement, positioning, and, in some cases, effectiveness of your team. Being able to determine the elevation and relief of areas on standard topo maps is a required skill for navigators. To do this, you must first understand how the mapmaker indicated the elevation and relief on the map.

Elevation of a point on the earth's surface is the vertical distance above or below mean sea level (MSL). Elevation above or below MSL is expressed in meters (m) or feet (ft).

Relief is the representation of the changes in elevation and the shape of the earth's surface (as depicted by the mapmaker). Relief depicts the shapes of hills, valleys, streams, and terrain features on the earth's surface. Contour lines are the tool the mapmaker uses as a common language to portray terrain features. Being able to read the relief on a flat map will allow you to visualize the actual terrain before you see it in person.

Altitude is the vertical distance above a reference level, which can be mean sea level or ground level depending on the context.

METHODS OF DEPICTING RELIEF

Mapmakers use several methods to depict relief of the terrain. The most common and the one we will focus on are *contour lines*. A contour line represents an imaginary line on the ground, above or below sea level that trace or connect points of equal elevation. All points on the contour line are at the same elevation. The elevation represented by contour lines is the vertical distance above or below sea level. The closer the contour lines are together the steeper the terrain is.

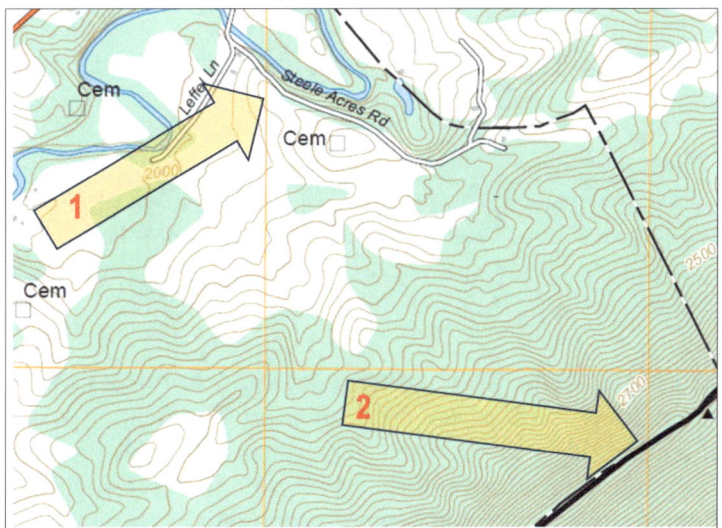

The closer together the contour lines, the steeper the terrain. Notice the density and number of contour lines along these two "routes". Walking along **1** will be fairly easy due to the flat terrain; walking along **2** will be more difficult due to the steep terrain.

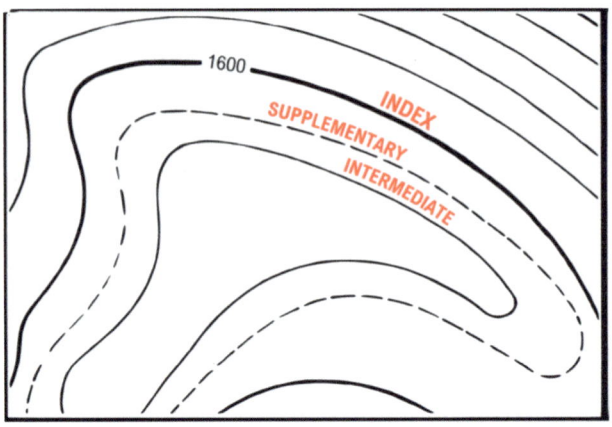

The three types of contour lines on a topo map

The three types of contour lines used on a standard topographic map are:

Index. Starting at zero elevation or mean sea level, every fifth contour line is a heavier line. These are known as index contour lines. Normally, each index contour line is numbered at some point. This number is the elevation of that line.

Intermediate. The between the index contour lines are called intermediate contour lines. These lines are finer and do not have their elevations given. There are normally four intermediate contour lines between index contour lines each representing one unit of the contour interval for the map. If the contour interval is 20 feet the elevation change between each line will be 20 feet.

Supplementary. These contour lines resemble dashes. They show changes in elevation of at least one-half the contour interval. These lines are normally found where there is very little change in elevation, such as on level terrain.

Shaded Relief. Some map printers use a technique of shaded relief that places shading on the map. This works along with the contours to make the features' elevation changes and shapes stand out. If you are new to using topo maps, be sure you don't confuse the vegetation on a standard topo map with this technique. Shaded relief maps are not standard, but they may be an option you prefer using. Use what works for you.

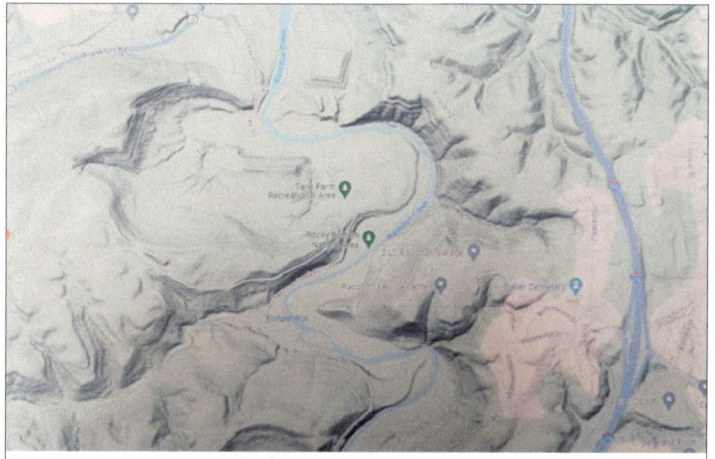

Shaded relief maps can help make the terrain elevation stand out. These are not the preferred nav map but if they work for you go for it. (photo courtesy of The Modern Minuteman, Jay Pallardy)

Contour Interval

Before the elevation of any point on the map can be determined, you must know the contour interval for the map. The contour interval may be in meters or in feet. The interval measurement is the vertical distance between adjacent contour lines.

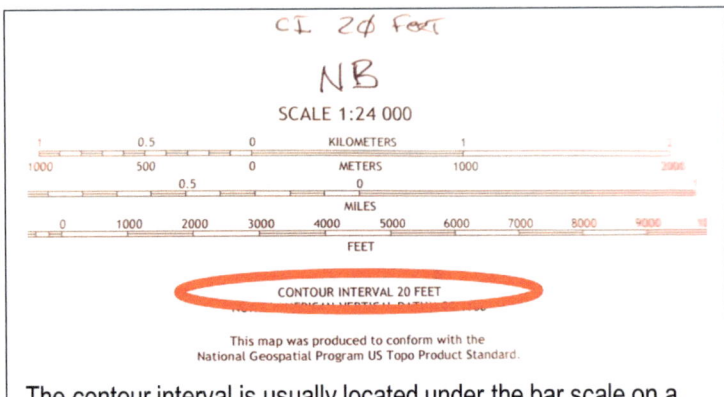

The contour interval is usually located under the bar scale on a topo map. The writing above the scale is mine, this is a photo of one of my maps with some of my typical margin modifications.

Determine the elevation of a point on a map

Use the contour interval and the unit of measure (feet or meters) and find the numbered index contour line nearest the point where you are trying to determine the elevation. If you are increasing elevation, add the contour interval to the nearest index contour line. If you are decreasing elevation, subtract the contour interval from the nearest index contour line.

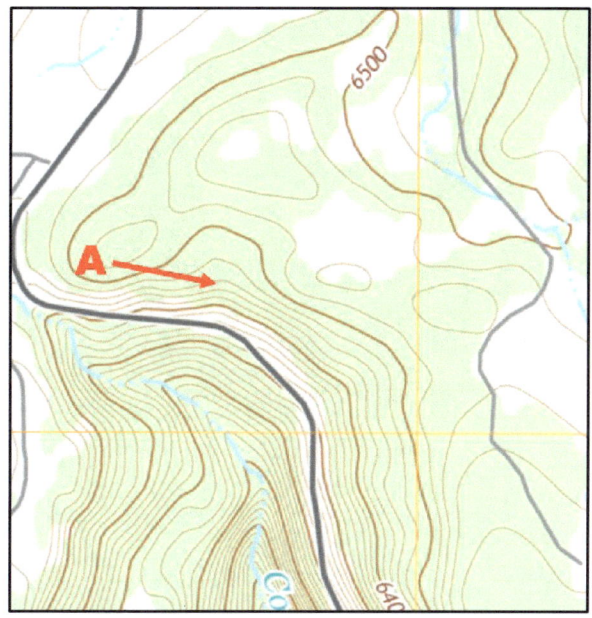

Determine if you are going from lower elevation to higher, or vice versa. Point A is between index contour lines, the lower index is numbered 6400, which means any point on that line is at an elevation of 6400 feet above mean sea level. The upper index contour line is numbered 6500, or 6500 feet. Going from the lower to the upper index contour line shows an increase in elevation.

To determine the exact elevation of point A, start at the closest index contour line (6500) and count the number of intermediate contour lines to point A. Point A is on the second intermediate contour line below the 6500- index contour line. The contour interval is 20 feet so each one of the intermediate contour lines crossed to get to point A subtracts 20 feet from the 6500-foot index contour line making point A at 6460 feet.

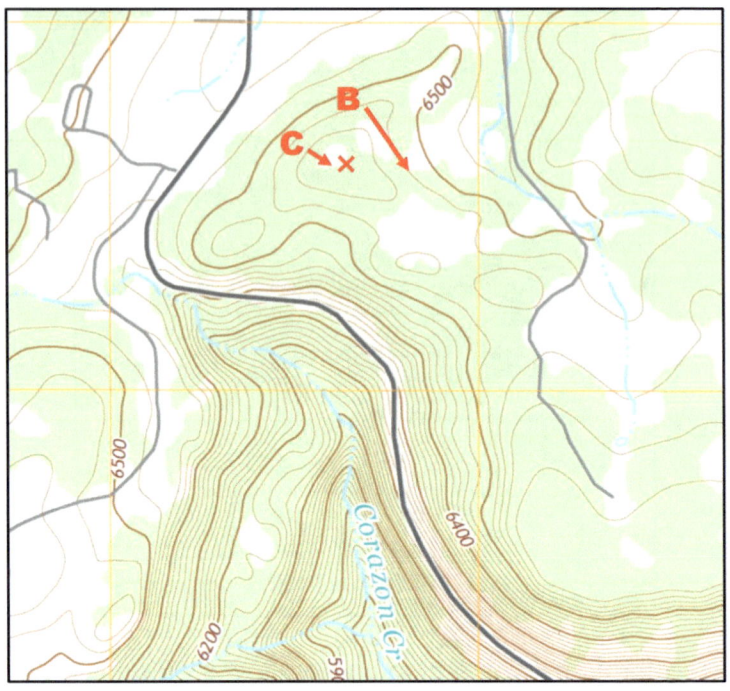

To determine the elevation of point **B**, find the nearest index contour line. In this case, it is the lower index contour line numbered 6500. Locate point **B** on the intermediate contour line immediately above the 6,500-meter index contour line ("above" means uphill or a higher elevation). Point **B** is located at an elevation of 6,520 feet.

To determine the elevation to a hilltop at point **C** add one-half the contour interval to the elevation of the last contour line. In this example, the last contour line before the hilltop is the 6540 contour. Add one-half the contour interval, 20 feet, to the index contour line to get the approximate elevation of the hilltop. The elevation of **C** in this example is 6,550 feet.

In addition to contour lines, map makers use benchmarks and spot elevations to indicate points of known elevations on the map.

Benchmarks are the more accurate of the two. They are signified by a black X, such as **X BM 3088**. The 3088 indicates that the center of the X is at an elevation of 3088 units of measure above mean sea level. There is a physical marker at each BM, it may be a concrete post with a plate or just a simple metal rod. These are reaching the end of their service life in many cases and may no longer be present due to environmental deterioration.

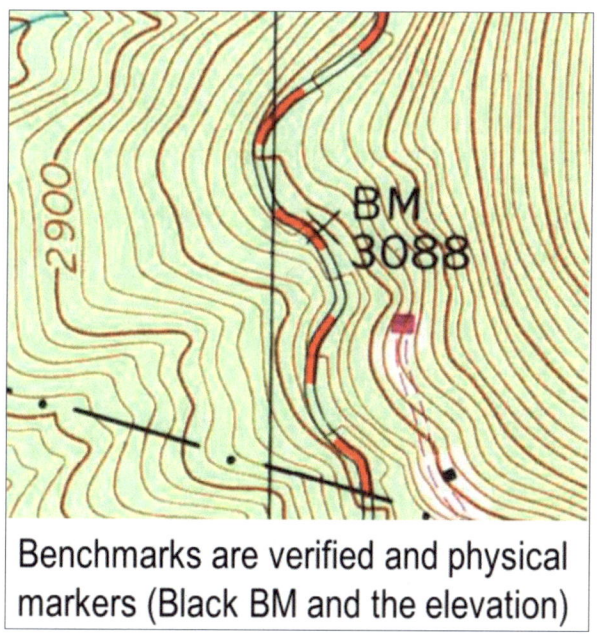

Benchmarks are verified and physical markers (Black BM and the elevation)

Spot elevations are shown by a brown X and are usually located at road junctions and on hilltops and other prominent terrain features. If the elevation is shown in black numerals, it has been checked for accuracy; if it is in brown, it has not been checked.

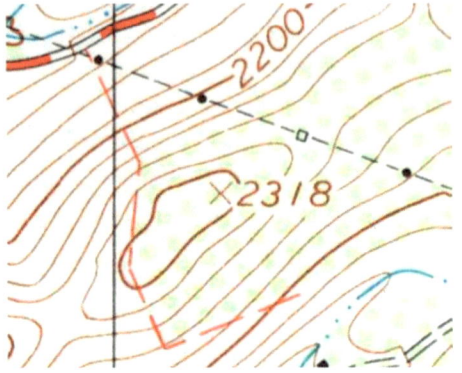

Unverified spot elevation (Brown X and text)

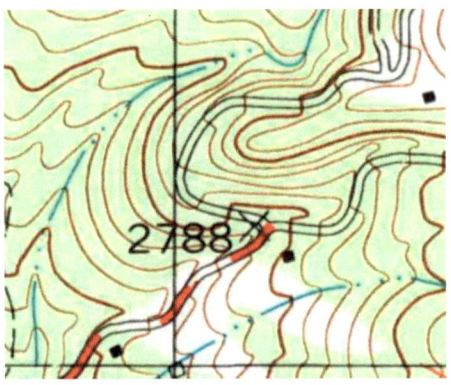

Verified spot elevation (Black X and text)

NOTE: Some new maps are being printed using a dot instead of brown Xs. The new series of USGS topo maps do not have benchmarks or spot elevations at all. Historical series USGS maps still have these and will follow the Benchmark annotation rules above.

SLOPE

As you start planning your navigation routes you will be interested in the slope of the terrain. Slope is defined as the rate of rise or fall of a terrain feature. What this translates to on the ground is trafficability of a leg or route, the aspects of terrain, and the effects on you (more on this in Chapter 3). And slope is an indication of the unmeasurable factor of physical fatigue and misery. Your speed along the route will be influenced by this factor as well; steep terrain will slow individual movement. This slope can be determined from the map by studying the contour lines—the closer the contour lines, the steeper the slope; the farther apart the contour lines, the gentler the slope. Keep in mind the contour interval also gets a vote; a 20 meter contour interval with lines close together will be much steeper than a 10 foot interval that is similarly spaced. The four types of slopes that concern us are:

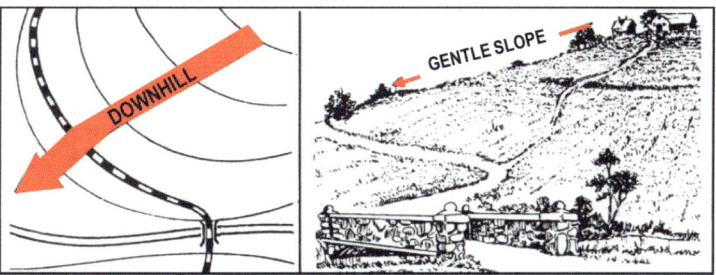

Gentle Slope. Contour lines showing a uniform, gentle slope will be evenly spaced and wide apart. Walking will be challenged slightly by gravity and vegetation, but it won't be too difficult.

87

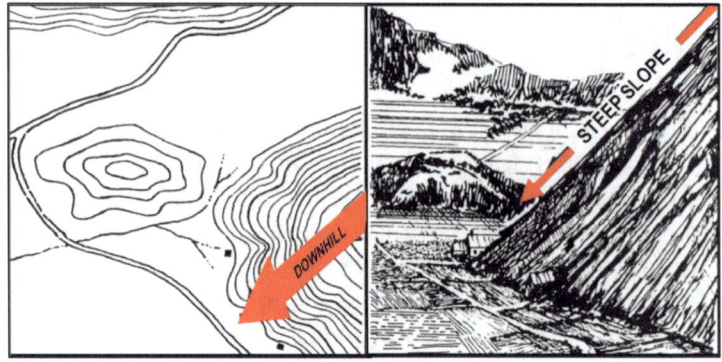

Uniform Steep Slope. Contour lines showing a uniform, steep slope on a map will be evenly spaced and close together. The closer the contour lines are to each other, the steeper the slope. Walking or maneuvering on this can present a significant physical challenge or even become impassable if wet weather changes the soil conditions.

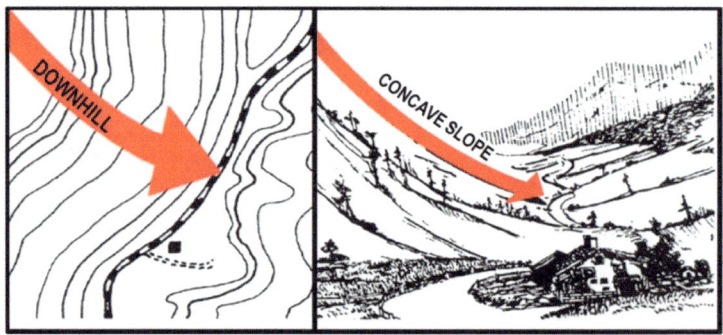

Concave Slope. Contour lines showing a concave slope on a map will be closely spaced at the top of the terrain feature and widely spaced at the bottom. The ascent will be easy at first, but the climb will become more difficult you get farther up the slope. Vegetation or tree covered terrain can conceal these slopes, a route that seems viable from a distance can actually be impassable due to the steepness of the final climb.

Convex Slope. Contour lines showing a convex slope on a map will be widely spaced at the top and closely spaced at the bottom You will have an easier climb as you near the top as it flattens out. Travel along the upper portion of a convex can result in skylining your movement due to the long gradual upper slope.

There are numerous other skills to learn about the aspects of terrain and contour. Vertical interval, line of sight calculations, ridgelining etc. These skills are useful, but for foundational land navigation and the applicable terrain analysis required for our community skillset, we will defer them to other manuals to keep our focus in here.

CHAPTER 3
PLAN THE ROUTE

This chapter covers the art and science of route planning. Route planning is more than just punching coordinates and waypoints into a GPS. You must account for the terrain, mission, and variables when you plan your routes. Visualizing the terrain's relationship to you/your team and any threats that can influence your route is the graduate level work.

METT-TC REVIEW

We will briefly review METT-TC (see the CM-1 and CM-2 manuals for a deeper discussion). These mission variables allow you to visualize yourself in relation to the environment, threats, and other friendlies and how they affect the route you are planning. The mission variables are Mission, Enemy, Terrain and Weather, Troops and support available, Time available, and Civil considerations (METT-TC). METT-TC can be looked at as overused or a "catch all" that has been over referenced by the tactical community. Do not fall into the trap of dismissing it because of this mis- and overuse.

Mission. Mission refers to the purpose (the main action you are taking) and the tasks you must accomplish. Assessing the mission includes understanding the *intent*. For example, is it more important to get there as fast as possible or to move undetected? Remember unless you are training navigation is not the primary task.

Enemy. The second component of METT-TC. You will account for opposing threats, including their strength, tactics, and weaknesses. This assessment will influence your route planning. You might avoid a section of a city or you may decide to time your route so you cross a piece of terrain at a specific time to avoid a threat. Remember the "E" in METT-TC is always the focus; the threat or enemy must always be the main thing.

Terrain and Weather. The third component of METT-TC is Terrain and Weather. This is the physical environment you will navigate through. Account for the impact of the environment on your mission, such as the effects of terrain or weather conditions on your mobility and visibility. Use the advantages of weather and darkness during route planning (terrain analysis is discussed later in this section using the acronym OAKOC).

Troops and Support Available. The fourth component of METT-TC is assessing the strengths and weaknesses of your teammates, family members, supporting or supported local friendly forces, including manpower, their capabilities, readiness, equipment, and supplies. Does your team have the physical ability to safely cross the six foot deep stream along your route, or do you need to navigate to a crossing site? Troops and support available is also knowing yourself and assessing current capabilities.

Time Available. The fifth component of METT-TC refers to the amount of time available to plan and execute the mission, including deadlines and the overall pace of operations. This includes the time required to move in relationship to the enemy and conditions, as well as available light if the mission requires movement only during hours of darkness.

Civil Considerations. The final component of METT-TC are the Civil considerations in your area. You must account for the demeanor and state of the local population during a widespread crisis. We must be aware and tuned in to local population sentiment and attitudes toward us, local authorities, and each other during widespread social unrest.

Another civil consideration for us is private property and trespassing. Plan your dismounted routes accordingly even during a crisis. If you must execute your dismounted get home plan from work or school, you had best account for this variable. Cutting across a farmer's field in broad daylight because it is convenient may not be good for your health and safety.

Do not wish away civil considerations during your route planning. Your personal mission and priorities don't mean a whole lot to other people - especially landowners.

INTERPRETATION OF TERRAIN FEATURES

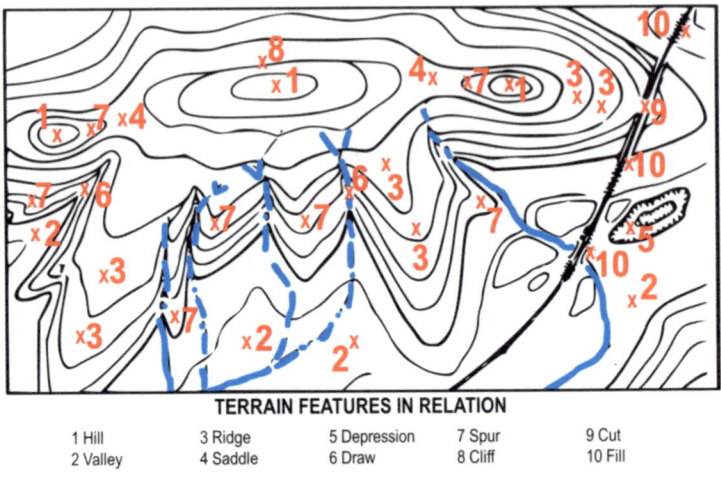

TERRAIN FEATURES IN RELATION

1 Hill	3 Ridge	5 Depression	7 Spur	9 Cut
2 Valley	4 Saddle	6 Draw	8 Cliff	10 Fill

Terrain features do not stand alone, it is the relationship on the ground that we must understand to become a proficient navigator and an effective small unit leader. Knowing how to read and analyze terrain depicted on a map will paint a picture for you, it allows you to visualize a route and how it will unfold as you walk across the actual ground. We need a system or framework to help put this together.

This terrain analysis (the terrain features in relation coupled with mission variables of METT-TC we reviewed earlier) is done by using the military aspects of terrain.

OAKOC (MILITARY ASPECTS OF TERRAIN)

We use the military aspects of terrain to analyze the ground and determine the effects of the terrain on us and any hostile groups. This analysis always has a purpose; it is not to recite the acronyms or build a checklist. This is an analysis done to provide information about your routes. Allocate as much time as possible to analyzing these aspects during your planning. That being said, you must also be able to rapidly analyze a piece of terrain as you move and quickly adjust your plan.

This analysis allows us to template the probable location of hostiles, position our own assets for security, select movement techniques and formations, and plan timing and location of combat multiplier (aka "force multiplier") employment (a drone for example).

OAKOC - **Obstacles.** An obstacle is anything that blocks or impedes your movement. Identify existing (either natural or man-made) and reinforcing obstacles in your area. When planning your route you must avoid or overcome/ bypass obstacles like rivers, swamps, cliffs, and manmade obstructions or events (mobs and riot roadblocks, fences etc). Obstacles are relative to how you are traveling. If you are driving, a locked gate across a national forest road is an obstacle; if you are walking it is not. You will know about some obstacles ahead of time either by your map, news, or reports from your team. Some obstacles will be a surprise to you such as roads flooding, bridges down, or hostile armed protester checkpoints. You must be able to adjust your route on the fly and quickly change your plan based on the new situation.

Existing obstacles are things that may impede movement (mounted or dismounted). Natural existing obstacles include rivers, dense forests, even tree stumps and large rocks (can impede off road vehicular movement). Existing obstacles of the man-made type include canals, railroad embankments, buildings, agricultural fencing etc.

Context also matters when assessing obstacles. Under normal conditions this is an existing obstacle for dismounted movement; in a true emergency some existing obstacles are only a slight inconvenience. (pro hillbilly tip, smooth single wires with insulators indicate they are "hot" or have electricity running through them.)

Reinforcing obstacles are deliberately emplaced to inhibit the ability of the opposing force to move, mass, and reinforce. Reinforcing obstacles are constructed, emplaced, or detonated by a military or guerilla / criminal force. Examples include hasty roadblock debris (and even people), ditches, abatis, burning tires, and concertina wire obstacles / emplaced fencing.

OAKOC - **Avenues of Approach** is the anticipated or templated routes or groups of routes. An Avenue of Approach* is the route leading to an objective or key terrain. Avenues of approach are classified by type (mounted, dismounted, air, or subterranean), formation, and speed of the largest unit (group) traveling on it.

friendly forces are not referred to as having Avenues of Approach; we use Routes or an Axis of Advance when referencing friendly maneuver graphic control measures after our routes have been planned.

OAKOC - **Key Terrain** is location(s) or areas whose control gives a marked advantage. It is a conclusion arrived at after analysis rather than a simple observation. A hilltop overlooking an avenue of approach may or may not be key terrain. Even if it offers clear observation and fields of fire, it means nothing if a threat can easily bypass it or maneuver on a different avenue of approach.

OAKOC - **Observation and Fields of Fire** are locations along each avenue of approach providing clear observation and fields of fire. Analyze the area surrounding key terrain, objectives, engagement areas, obstacles and locate intervisibility lines (ridges or horizons which can hide equipment or personnel from observation). Identify positions where observers can direct assets against your team. Consider this from both threat and friendly perspectives. Identify the dead space in your area of operations to determine where you are vulnerable along your route.

OAKOC - **Cover and Concealment** is mistakenly confused by many; they are not interchangeable terms. Cover is natural or man-made ballistic protection from the effects of direct and indirect fire. Cover may change and become ineffective at any time as new weapons systems are brought to bear against it. Concealment is protection from observation and provides no relevant ballistic protection. Consider the terrain, vegetation, structures, and other features along routes (and on objectives or key terrain) to identify sites offering good or poor cover and concealment. Look at it from both friendly and threat perspectives; determine who has the advantage and why.

Categories of terrain

When analyzing terrain for ground maneuverability during the IPB Process (Intelligence Preparation of the Battlefield or Intelligence Preparation of the Battlespace) the terrain is placed into one of three categories. These indicate the difficulty of maneuver through a series of terrain. *Unrestricted* terrain is easy, *restricted* will slow your movement (steep terrain, or passable swamp), and *severely restricted* is near impossible without some type of engineer, watercraft, or air mobility assistance (cliffs, flooded areas etc.) Assessing these during the IPB process is not addressed in this manual but you should become familiar with them since all these concepts work together. You must understand the reality of the terrain and the effects on your movement when you are looking at a topo map and planning a route.

TERRAIN RELATED TERMS

Some of these terms are doctrinal, some are unofficial / institutional knowledge but are part of our common language.

Go, Slow-Go, No-Go (Terrain). In military terms, "go", "no-go" and "slow-go" terrain were used to describe the maneuverability of the terrain. These have been replaced with *"Unrestricted, Restricted, and Severely Restricted"* however you may still hear the older terms (incorrectly) used.

High Ground. Terrain that is higher in relation to surrounding ground. High ground may or may not provide a tactical advantage.

Low Ground. Terrain that is lower / has less elevation in relation to the surrounding terrain. Low ground typically has denser vegetation, swampland, or water and can be (not always) a tactical disadvantage.

Bowl. Low ground surrounded by higher ground on all sides. It is not a depression per se, it may be a portion of a valley. They can be large and wide open or wooded and small. They are usually not a good piece of terrain from a tactical perspective.

Wadi or Arroyo. A dry desert riverbed that will catch and channel water during rainy seasons. They can flood unexpectedly with no signs of local rain so use caution when overnighting in or near them. They may provide you with good covered and concealed routes, but they are also natural lines of drift.

Intervisibility Line (IV Line). Intervisibility Lines (IVLs) are *not* terrain features and do not appear on maps as discrete entities. IVLs are an effect of terrain on the line of sight / observation. Terrain features can be readily identified regardless of the observer's location, IVLs are only relative to the observer's perspective (elevation and direction). IV lines are a localized pattern of limitations on observation caused by subtle variations in terrain elevation (aka folds in the ground). Picture a long, straight, two lane road. Oncoming vehicles may seem to disappear in spots as they approach. These are IV Lines. IV lines can provide a significant advantage to the side who knows the terrain; a proficient unit or team can infiltrate laterally along concealed IV lines.

Note how the lower third or so of the fence posts "disappear" in this IV line. The contour interval of the map does not account for this small fold in the ground, so it does not show up as a draw or valley. Depending on the map, the contour interval, and the terrain there may be unseen opportunities to travel undetected.

Danger Area. The term "danger area" refers to any area on the route where the terrain would expose you or your team to enemy observation, fire, or both. Examples include large open areas, roads, trails, and bridges or crossing sites over water obstacles. As you plan your routes account for these and be deliberate where you cross or bypass. (see CM-2 Recon manual)

Linear Danger Area. A linear danger area is an area where the team or unit's flanks are exposed along a relatively narrow field of fire. Examples include building lined streets, tree lined roads, narrow trails, and streams.

High Speed Avenue (of Approach). An unrestricted mobility corridor that affords a mounted enemy access to or through a piece of terrain. Hardball roads with little restrictive terrain or canalization are typically high-speed avenues of approach.

Military Crest. An area on the forward or reverse slope of a hill or ridge just below the topographical crest. Traveling below the military crest will prevent skylining as you move (silhouetting yourself against the sky to an observer)

Hardball / Hardball Road. A paved road.

Goat Trail. A very narrow trail through rough or difficult terrain, usually (but not always) single track. Some will call any tank trail or dirt road a goat trail, but this is incorrect even though this is not an "official" term.

Defile. A defile is a narrow pass or gorge between mountains or hills. The term originates from a military description of a route through which troops can march only in a narrow column or with a narrow front. A defile may be defined by a narrow valley, draw, or cut (all terrain features) or by vegetation, manmade structures or geographic features that may not be captured on your map as terrain features.

Single Track. A path or trail wide enough for a single walker, mountain bike, or motorcycle. Single track may not be passable by all means; terrain and weather will dictate the trafficability.

Single track routes are just wide enough for single file walking, a mountain bike, or a dirt bike.

ROUTE SELECTION

Route selection is choosing the route you want to travel along; this includes both vehicle and dismounted movement. When we say route selection it isn't just choosing a forest service trail or a road / interstate number. A "route" contains all the segments you put together for your trip or mission while accounting for the METT-TC / OAKOC factors. During a crisis you may choose a route to avoid people (depending on the situation) or you may prioritize speed if the roads are clear and you need to get to your destination quickly. Take the entire situation into account; your mission (what you are setting out to do), any threats that may be present, the terrain, who is traveling with you, the time you have available to make the movement, and what the local population is like at the time.

Do you want to be seen or not? This is one of the simple questions or factors to consider when choosing your route. You must assess the situation to figure out what the risk is. Human contact may not be desired at all if the situation has deteriorated, this factor alone will influence your route and even your method of travel.

You may choose to travel only at night or in inclement weather to conceal your movement. You may want to avoid roads and trails. If you are trying to get help as quickly as possible you may want to stick to main roads and open ground in the hopes of being seen. The bottom line is your mission will dictate your route choice.

ROUTE PLANNING

These principles apply to moving as an individual, two or three person teams, or movement with tactical formations. OAKOC always applies; as you develop your skills you will be able to visualize your route over the terrain and how the aspects of terrain change as you move. It is a blend of art and science, you will begin to mentally traverse the route based on how you see the topo map "lets see...contour lines close together, will be steeper and slow movement but the underbrush will be less, and it will most likely be dry; Ill follow the ridgeline north about a third of the way parallel to the ridge to avoid the low ground" etc.

Use Terrain for Protection. Terrain can offer cover and concealment. Dismounted formations and techniques of movement (see CM-1 and 2) can help you use the terrain to your advantage. Choosing routes that keep a major terrain feature between you and a potential threat can be useful. If you want to remain unobserved avoid "skylining" or moving near the top of a ridge or hill while silhouetting yourself. Staying below the military crest of a hill, spur, or ridge can protect you or a team from observation.

Travel time. Estimating the time for the route is a bit more art than science. You need to know yourself, your team, and the effects of the conditions, your load, terrain and vegetation on your rate of movement for the potential route (METT-TC and OAKOC yet again).

A general guide we can use is Naismith's rule. Naismith's rule can inform route planning by helping you calculate how long it will take to travel the route. including any extra time taken when walking uphill.

This rule of thumb was devised in 1892 by William Naismith, a Scottish mountaineer. This is just an estimate or guide, it is not really a "rule". The guideline is 5km an hour plus 10 minutes for each 100m of rise in elevation. It works out to 12 mins per kilometer plus the rise adjustment. Recommend you do not use this at face value for route planning, just use it as a *guide* during training iterations. Naismith's rule is considered the minimum time to complete a route for the slowest person in the team.

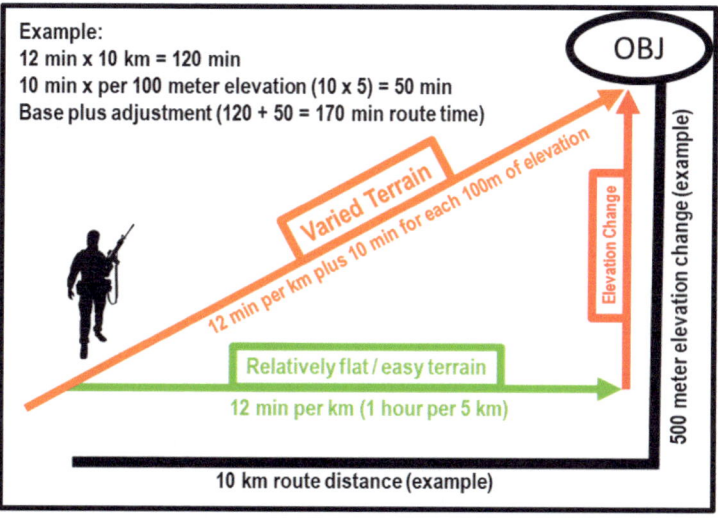

Example:
12 min x 10 km = 120 min
10 min x per 100 meter elevation (10 x 5) = 50 min
Base plus adjustment (120 + 50 = 170 min route time)

OBJ

Varied Terrain
12 min per km plus 10 min for each 100m of elevation

Elevation Change

500 meter elevation change (example)

Relatively flat / easy terrain
12 min per km (1 hour per 5 km)

10 km route distance (example)

Naismith's rule is just a start point for your route time planning. Collect your own data during training so you have a set of refined and tailored planning factors.

Record the actual rates / numbers from your routes and legs during training and use this tailored data for your next outing. There are corrections you can look up for the rule, but these can quickly turn into mathematical, second-guessing rabbit holes.

There is another TTP that can help get your initial numbers, it is the 10 percent rule of thumb. For every added condition that can slow movement such as broken ground, heavy loads, night, etc add ten percent to your usual rate. The learning point is you should keep notes to help develop your own set of planning factors to use for route planning.

Travel distance. The most direct route may be the shortest, but we need to consider all the factors we discussed earlier. Measuring the distance of potential routes is an easy task, but it can be tedious.

Features and checkpoints. Plan your routes for success, make it as easy as you can by planning around identifiable features. This will help you stay on your route. If you are planning a route for a team you must plan for rally points and lateral movement along each of your routes; recognizable features will facilitate these. Some of these attributes will be obvious when looking at your map, some will only come with experience.

Linear features that cross the route make excellent checkpoints or rally points. Examples include perennial streams, hardball roads, railroads, and power lines.

Checkpoints located where changes in direction are made are critical. If possible, it is always a best practice to plan your turns so they are recognizable even when dead reckoning. At a minimum know where your backstops are in case you miss an azimuth change at night. The backstop will let you correct your movement before it gets away from you. Missing a direction change at night with a tired team following behind you is a nightmare for everyone. Checkpoints are a way to help keep any mistakes small.

Dog Leg. This is an intentionally planned hard angle turn near the end of a route to provide a tactical advantage for the team. Changing direction before the final movement to a navigation objective vs going "straight in" is a best practice to prevent a hostile force from directly following a team.

Offset. A deliberate offset is a planned magnetic deviation to the right or left of an azimuth to an objective. In civilian nav training you may hear this called "aim off". Use it when the objective is located along or in the vicinity of a linear feature such as a ridge, road or stream. Because of errors in the compass or in map reading, the linear feature may be reached without knowing whether the objective lies to the right or left. A deliberate offset by a known number of degrees in a known direction compensates for errors and ensures you know whether to go right or left to reach the objective. Ten degrees is an adequate offset for most tactical uses. Each degree offset moves the course about 18 meters to the right or left for each 1,000 meters traveled. For example, in the figure below, the number of degrees offset is 10; so for a 500 meter leg the offset distance will end up being 90 meters.

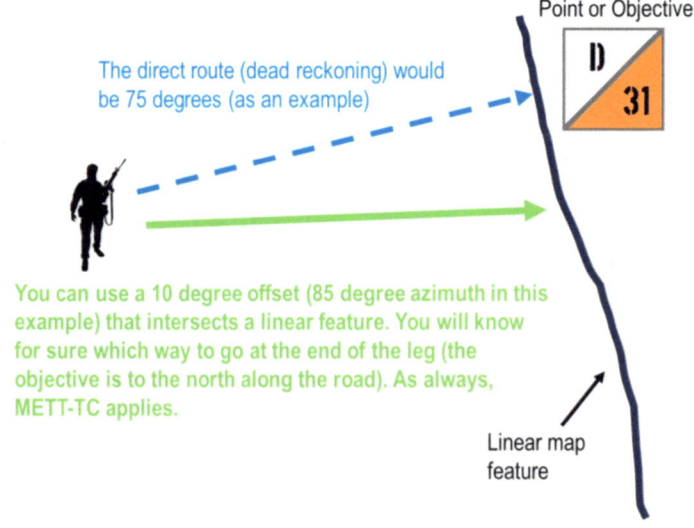

The direct route (dead reckoning) would be 75 degrees (as an example)

Point or Objective

D 31

You can use a 10 degree offset (85 degree azimuth in this example) that intersects a linear feature. You will know for sure which way to go at the end of the leg (the objective is to the north along the road). As always, METT-TC applies.

Linear map feature

Catching Feature or a Backstop. A backstop is an easily identifiable terrain or manmade feature that will indicate you have gone too far beyond your destination or checkpoint. Linear features like roads and streams make great backstops, these are "when I reach it, I know I have gone too far."

Attack Point. A recognizable feature a few hundred meters from your end point / objective that you can easily find. You will use the attack point to "reset" your nav route, they can assist with finding more difficult objectives that may not be near an identifiable feature. Once you reach your attack point double check your location and then execute the final short leg you had already planned. This shorter (and more manageable) leg will bring you to your final objective.

Handrail. A handrail is similar to a backstop, but they run parallel or partially parallel to your desired route. Handrails are linear features like roads or highways, railroads, power transmission lines, ridgelines, or streams that run roughly parallel to your direction of travel. Crossing a handrail or losing track of it will indicate lateral departure from your route. These can also be used (if the tactical situation dictates) to guide navigation by walking to the side but within visual of the feature. Again, linear features like roads and streams work best, just use caution with the natural lines of drift and assess the threat risk using the METT-TC and OAKOC tools. One admin caution on handrails, many military courses and selections will disqualify you for handrailing too close to a feature, especially if it is a road or tank trail.

Legs. The best way to manage a route is to divide it into segments called "legs." By breaking the overall route into several smaller segments, you can manage the longer route during your movement. Legs have only one distance and direction; any planned change in direction will end the leg and begin a new one.

A leg must have a defined beginning and end, marked with a graphic control measure such as a checkpoint or phase line. (When using GPS, these are captured as waypoints.) When possible, the start point and end point should correspond to a recognizable navigational aid (catching feature or navigational attack point).

Navigational legs (if executed properly) can reduce navigation errors over long distances. The normal errors / tolerance stacking over a single long dead reckoning (point navigation) route can add up and induce significant issues. For example, an error of just 5 degrees will result in missing an objective only 1k away by approximately 87 meters. Most compasses are guaranteed in the 2-3 degree range of accuracy per manufacturer. Inaccurate readings, drift, and difficult terrain all add up to compound the issue. With shorter legs it lessens the opportunity for these errors to stack up on you and cause a significant navigation mistake.

To develop a leg first determine the type of navigation and route that best suits the situation. Once these two decisions are made, select the distance and direction from the start point to the end point and identify critical METT-TC information as it relates to the specific leg. If time allows it can help you visualize the route by drawing a sketch on a route chart or route card (an example is in the diagram below). Do not take this sketch on a tactical mission, this is simply a visualization, planning, and rehearsal tool you can use.

LEG	AZIMUTH / DISTANCE	KEY INFORMATION	
Leg 1: SP 1 to CKP 1. - Stay in woodline east of the hardball at the base of the hill.	008°/500m	O: Limited. A: HWY 1. K: Hill mass west of leg. O: N/A. C: Good.	
Leg 2: CKP 1 to CKP 2. - Stay on south side of dirt secondary road. Continue movement to the large church at CKP 2.	030°/1800m	O: Unlimited. A: Dirt trail. K: Hill 18 southeast of leg. O: N/A. C: None.	
Leg 3: CKP 2 to RP. - Stay on east side of hill 25. Continue movement to the boulders at the RP.	319°/450m	O: Unlimited. A: Dirt trail. K: Hill 25 west of leg. O: N/A. C: None.	

LEGEND

CKP	CHECK POINT	SP	START POINT
RP	RELEASE POINT	M	METERS

111

DISMOUNTED NAVIGATION IN A CRISIS

Land navigation is always in context; it is done for a purpose. In our context it may involve movement during an emergency or widespread crisis. This movement to your safe area or back home makes you vulnerable. If you are alone and your mission is to get back home during a social crisis it is usually better to move unseen or unnoticed. When moving under sketchy conditions be sure to avoid roads and trails during the day since people will flock to the easy paths and routes. *There are no absolutes; your situation and the risk will guide how you proceed.* If you are healthy (uninjured) and have plenty of food and water, it is generally best to avoid human contact as you travel during a crisis. People are unpredictable...and crazy.

Your movement may be slow and deliberate or as fast as possible depending on the situation (I know...I keep saying situation dictates, but it is true. All factors combined will determine how you move). If there are abundant threats the more deliberate you are during movement the better. The greatest indicators of human presence are movement and sound. The smoother and quieter you are when walking the greater the odds you will see another person before they see you. Make frequent SLLS halts (Stop, Look, Listen, Smell). In daylight, scan and observe sections of your route before you move along it. Learn how to use the terrain and vegetation to your advantage. Do everything you can to remain concealed during a social crisis.

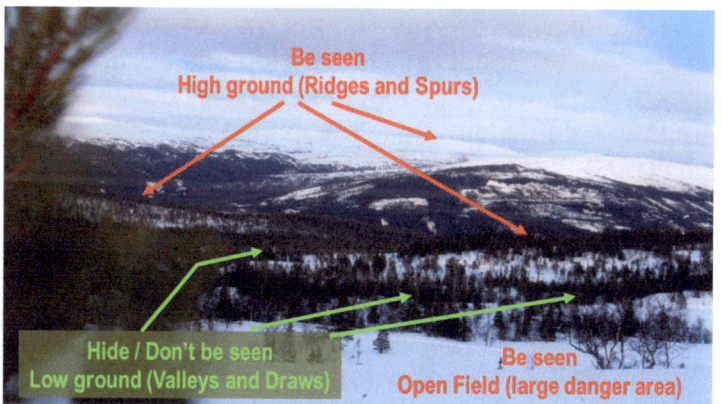

If you must move cross country during a crisis odds are you don't want to be seen. If you are lost and need to be rescued the opposite is true. Learn how to use the terrain to your advantage for your situation.

Choosing a route to minimize your exposure to human observation. The Route 2 option keeps you out of the open danger area (the field) and uses the low ground (valleys or draws) between the high ground (hills, ridges, and spurs).

113

Natural Lines of Drift. Natural lines of drift (NLOD) are the path of least resistance. Humans are lazy by nature and will move along the easiest routes. Roads, trails, low cut fields...anything that makes walking easier. Look at any college campus; the paths worn across the grounds are an excellent example. Humans are going to find the most direct, easiest path and take it. A key to evading human contact is staying off these paths and trails. There doesn't necessarily have to be a defined path or a trail, by looking at the terrain you will see the obvious routes across the ground. Some of these are apparent on the map, some only reveal themselves while you are moving along the route.

When planning your routes learn to identify and avoid the obvious NLODs; this is critical if you want to remain hidden. Sometimes paths and trails are unavoidable, but knowing the risks will inform a deliberate decision.

If you are trying to avoid contact stay off natural lines of drift, they attract humans. They don't always have to be paths or trails, they are just the easiest way to cross a piece of ground. Gaps in vegetation, open spaces, game trails (the paths that animals take through the terrain like this photo) must be avoided when threats are present.

Avoid Possible Kill Zones. Avoid large open areas surrounded by cover and concealment or those dominated by a piece of terrain that a threat would likely use. Watch for the presence of obstacles, traps, or any other signs of an engagement area (ground that a threat will likely want an opposing unit to be canalized into). If you are navigating to be rescued in normal social times the opposite is true, you will want to go where you will be seen. METT-TC applies no matter the conditions.

Move During Limited Visibility. Movement during darkness or other limited visibility conditions may provide concealment from threats. Combining a route's characteristics with the advantages of poor weather or limited visibility advantages during night can work in your favor. With the proliferation of NV and thermal devices this is not a complete solution, but a night movement can be the better option if there are hostile forces involved.

Obstacles. During a crisis the existing obstacles we discussed earlier are going to be an issue for us. Chain link fences, farm fences, and even entire urban areas can become an obstacle. Many of these are in the "just because you can it doesn't mean you should" category. Weigh the risks of a direct route that crosses a stream in cold weather or the direct route that requires you to scale a fence and cut through a farm field.

CHAPTER 4
NAVIGATE/STAY ON THE ROUTE
AND
RECOGNIZE THE OBJECTIVE

This is where it all comes together. This chapter discusses the application of all the technical skills to successfully navigate across terrain. Your route may be a series of dead reckoning legs with a distance and a direction or it may be a route that follows and flows through the natural features as you use terrain association.

USING DIRECTION AND AZIMUTHS TO NAVIGATE

Direction. Direction is expressed as units of angular measure. The method we use to express direction that is accurate, adaptable to any part of the world and has a common unit of measure is the degree. A degree divides latitude and longitude. You may see or hear Mil or Grad as units of measure as well, but for our purposes we will use degrees.

Maps do not have "up" or "down", they have cardinal directions (at a minimum) or more refined directions expressed in degrees.

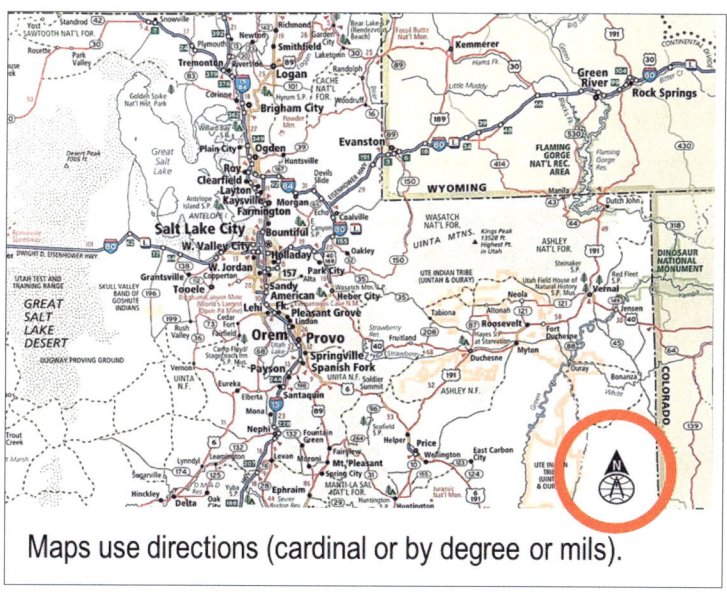

Maps use directions (cardinal or by degree or mils).

Cardinal Directions. Cardinal directions are the four main points of a compass and are north, east, south, and west (N,E,S,W). They are also known as cardinal points. On a compass rose (a graphic found on maps that shows directions), you'll see these four cardinal points. North is usually emphasized as maps are printed with North to the topo of the page. (North is never "up" or the way the map is facing, true north is determined by which direction the North Pole is and magnetic north is where a compass north seeking arrow points.

Cardinal Directions

Ordinal Directions. Ordinal directions (also known as intercardinal directions) refer to the directions located equally between each cardinal direction. The four ordinal directions are Northeast (NE) halfway between North and East; Southeast (SE) halfway between South and East; Southwest (SW) halfway between South and West; and Northwest (NW) halfway between North and West.

Ordinal Directions

Secondary Intercardinal Directions are between each cardinal and ordinal direction. Examples include NNW, NNE, and SSW.

Using cardinal or ordinal directions is not precise, they only provide a general orientation or direction. We might use these when conveying travel instructions to a family member or when sending a contact report. For example, you might tell your brother "I am along Ruby Creek where it crosses Route 45. From where you are just walk southwest until you reach Route 45 and then go west 100 meters until you reach the creek crossing." Or "head west on I-40 from Nashville to Exit 182 then go north on Highway 96".

Cardinal Direction	Degrees
North	0
East	90
South	180
West	270
Ordinal	
Northeast	45
Southeast	135
Southwest	225
Northwest	315

Cardinal and Ordinal directions expressed in degrees. Remember using cardinal and ordinal are not *usually* meant to be precise, they are general orientations. This is just general knowledge that can assist you with understanding how all these components works together.

ALL THE NORTHS

Base Lines. To express direction as a unit of angular measure, there must be a starting point or zero measure and a point of reference. These two points designate the base or reference line.

There are three base lines for maps: true north, magnetic north, and grid north. The two we will use are magnetic and grid north.

True North. A line from any point on the earth's surface to the north pole. All lines of longitude are true north lines. True north is usually represented by a star on a map.

Magnetic North. The direction to the north *magnetic* pole, as indicated by the north seeking needle of a magnetic compass. Magnetic north is not the north pole, it is the earth's strongest magnetic field. The other complicating factor is it moves over time, so what was magnetic north last year will be different this year. Magnetic north is symbolized by a line ending with half of an arrowhead. Readings are obtained with magnetic compasses.

Grid North. Grid North is established by using the vertical grid lines on the map. Grid north may be symbolized by the letters GN or the letter "y" on the declination diagram. Grid north is the northern direction along a map sheet easting. It is not referred to as "up".

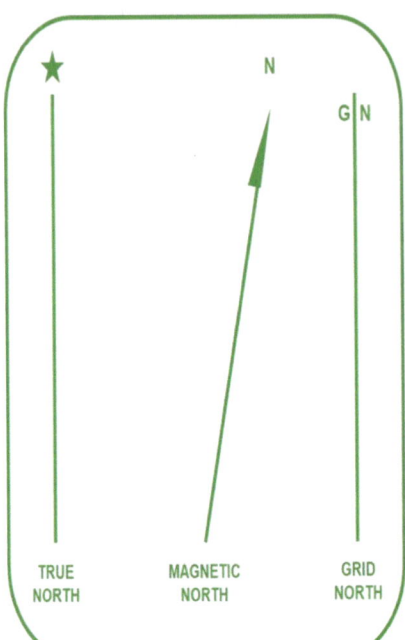

True North
(North Pole, also called
geodetic north
or geographic north)

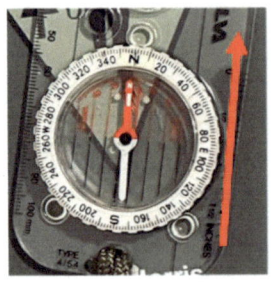

Magnetic North

Grid
North

Azimuth. An azimuth is defined as a horizontal angle measured clockwise from a north base line. Your north base line could be magnetic north (angle measured with a compass) or grid north (angle measured from an easting on a map). Azimuth is the most common method to express direction for US land navigation. When using an azimuth, the point where the azimuth originates is the center of an imaginary circle. This circle, just like any other is divided into 360 degrees.

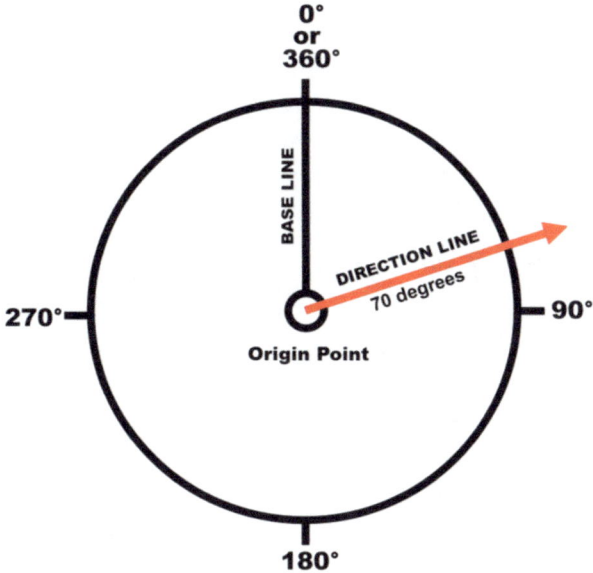

Back Azimuth. A back azimuth is the opposite direction of an azimuth and always 180 degrees difference. To obtain a back azimuth, add 180 degrees if the azimuth is 180 degrees or less and subtract 180 degrees if the azimuth is 180 degrees or more.

Grid Azimuth. When an azimuth is plotted on a map between a start point and end point, they are joined together by a straight line. Use your protractor to measure the angle between grid north and the line you drew; this measured azimuth on the map is the **grid azimuth**. When measuring azimuths on a map, remember that you are measuring from the start point to the end point. If you mistakenly take the reading from the end point (your objective) back to the start point, the grid azimuth (GAZ) will be opposite (a back azimuth) which will cause you or your team to go in the wrong direction. Always use your protractor, not your compass to measure on your map.

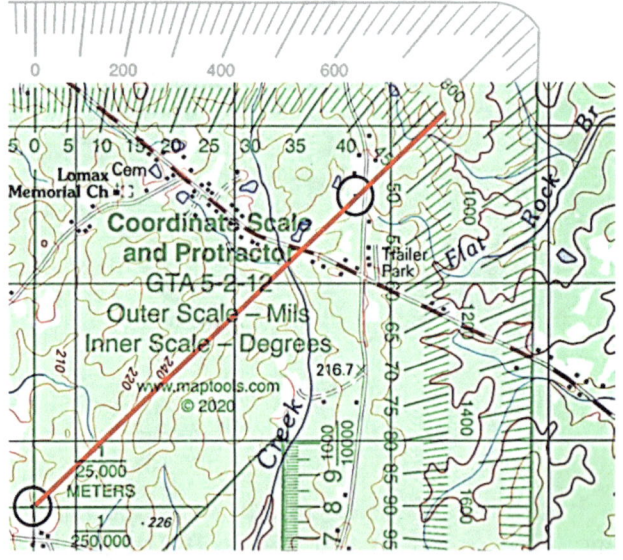

A grid Azimuth (GAZ) of 46 degrees on a 1:50,000 MGRS map.

AZIMUTH CONVERSION

Compass to Map and Map to Compass. To account for the difference between grid north (map north) and magnetic north (compass north) you must add or subtract. The trick is knowing when to add and when to subtract; other than that, the math is straightforward. A current declination diagram will give us the known values to add or subtract to our azimuths.

Declination Diagram. Declination is the angular difference between the norths. The most important one for us is the angle between magnetic and grid north. Your map is laid out in the projection that uses "Grid North" while your compass will point to the earth's magnetic north. Magnetic north changes each year and will shift considerably over time. It is important you know the date of the data on your map to ensure you are using relatively new declination diagram data.

The declination diagram shows the angular relationship, represented by prongs, among grid, magnetic, and true norths. While the relative positions of the prongs are correct, they are not drawn or plotted to scale. **Do not use the diagram drawing to physically measure the angle value.** This value will be written in the map margin (in both degrees and mils) beside the diagram. True North on the diagram is of little interest to us at this point in our progression so just ignore it for now.

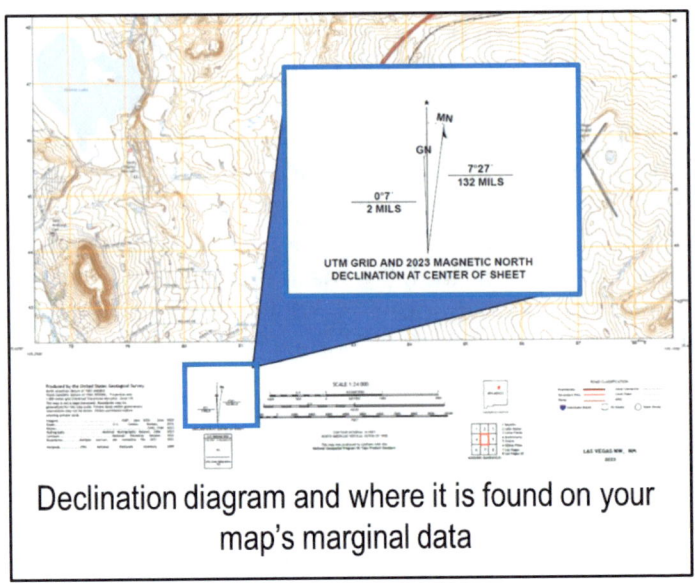

Declination diagram and where it is found on your map's marginal data

Grid-Magnetic Angle. The G-M angle value is the angular size that exists between grid north and magnetic north. This value is expressed to the nearest 1/2 degree or in degrees and minutes, with mil equivalents shown to the nearest 10 mils (again we will deal with degrees only for now). The G-M angle is important to us because azimuths translated between map and ground will be in error by the size of the declination angle if not adjusted for it.

Conversion. Since the location of magnetic north does not correspond exactly with the grid north lines on your map, a conversion from magnetic to grid or vice versa is needed. This is done using the G-M Angle found in the declination diagram in the marginal data. This is where we have to do some simple math to go from Magnetic Azimuth (MAZ) to Grid Azimuth (GAZ) or grid to magnetic.

126

Diagrams with Notes. This refers to the conversion notes that appear in conjunction with the declination diagram explaining the use of the G-M angle. The conversion (add or subtract) is governed by the direction of the magnetic-north prong relative to that of the north-grid prong. *USGS will rarely (if ever) have the conversion notes on the declination diagram, it is wise to write the formula in when you receive a new map.*

Your declination diagram may be incorrect. Due to the magnetic pole progressive movement the declination diagram printed on your map may be out of date. You can (and should) use the NOAA website to update the GM angle on your map sheets.

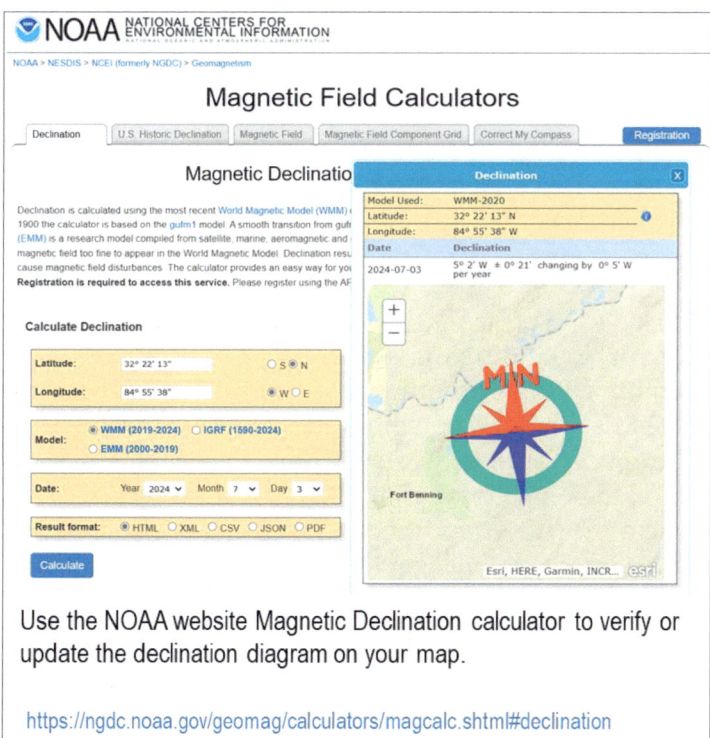

Use the NOAA website Magnetic Declination calculator to verify or update the declination diagram on your map.

https://ngdc.noaa.gov/geomag/calculators/magcalc.shtml#declination

Converting from One Type Azimuth to Another. The Western US has East Declination (Mag North Arrow is to the right (East) of the Grid North line in the Declination Diagram) while the Eastern US has a West Declination (Mag North Arrow is to the left (West) of the Grid North line in the Declination Diagram).

Magnetic Declination in the Continental US

Note the angles and severity of angle between grid and magnetic north at each coast. This is in a constant state of change as the earth's magnetic pole moves.
2020 data, source from NOAA.gov chart

Math Note. Remember, there are no negative azimuths. If the azimuth you are converting is less than the G-M angle (and you are using one of the two subtraction formulas) you will have to add 360 before you do the conversion math. Since 0 degrees is the same as 360 degrees, then 7 degrees is the same as 367 degrees to allow you to do the math. 7 degrees and 367 degrees are located at the same point on the azimuth circle. The small number azimuth can now be converted after adding 360 because it is now larger than the G-M angle (use only when subtracting).

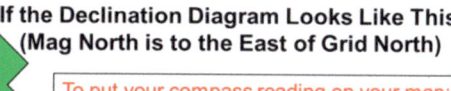

If the Declination Diagram Looks Like This
(Mag North is to the East of Grid North)

G N

MAGNETIC NORTH

To put your compass reading on your map:

You are converting a Magnetic Azimuth to a Grid Azimuth

You will **ADD** the G-M Angle to the Magnetic Azimuth

MAZ to GAZ → Add

To use your map reading on your compass:

You are converting a Grid Azimuth to a Magnetic Azimuth

You will **SUBTRACT** the G-M Angle to the Grid Azimuth

GAZ to MAZ → Subtract

If the Declination Diagram Looks Like This
(Mag North is to the West of Grid North)

G N

MAGNETIC NORTH

To put your compass reading on your map:

You are converting a Magnetic Azimuth to a Grid Azimuth

You will **SUBTRACT** the G-M Angle to the Magnetic Azimuth

MAZ to GAZ → Subtract

To use your map reading on your compass:

You are converting a Grid Azimuth to a Magnetic Azimuth

You will **ADD** the G-M Angle to the Grid Azimuth

GAZ to MAZ → Add

LARS rule. There are several mnemonic devices to remember these conversions, one of the common we use is the *LARS rule*. This stands for Left Add, Right Subtract (the text colors are to help illustrate the direction on the diagram below). It is a simplified method of the previous two diagrams to recall whether to add or subtract based on which direction you go on the declination diagram. It works no matter if it is an easterly or westerly declination.

If you are going left across the declination diagram you are adding the GM angle (LEFT, ADD). If you are converting from left side to right across the declination diagram you subtract (RIGHT, SUBTRACT) the GM angle. It is that easy.

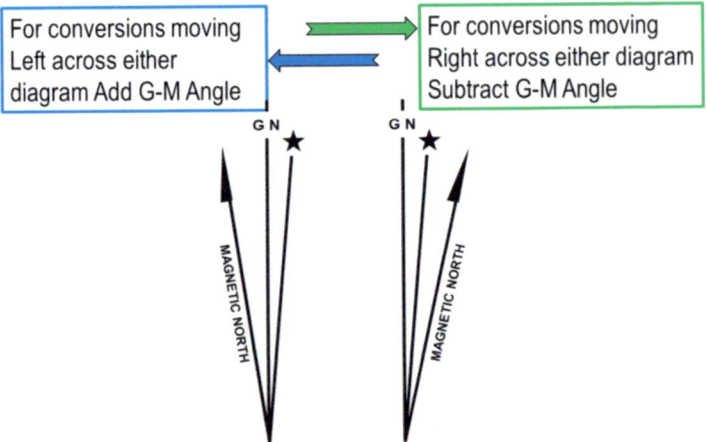

LARS Rule – The Left **A**dd, **R**ight **S**ubtract rule works the same way for both westerly and easterly declinations.

USING YOUR PROCTRATOR

Square Your Protractor. The protractor must be aligned parallel and perpendicular to the grid lines on the map when you use it. This is called keeping the protractor "squared up" and is required to take accurate readings and correctly plot locations and directions. When using your protractor, the base line is always oriented parallel to an Easting (a north-south grid line). The 0-degree mark is always toward grid north at the top of the map and the 90° protractor mark is to the east when measuring direction. Some multiple scale protractors are oriented differently, just ensure it is oriented properly for what you are measuring.

Your protractor will often not be directly centered on a grid line or intersection when plotting azimuths, so we need a technique to check alignment. Note the protractor is bisected at both the top and bottom by the same north-south grid line (Easting). Count the number of degrees from the 0-degree mark at the top of the protractor to this grid line and then count the number of degrees from the 180-degree mark at the bottom of the protractor to the same Easting. If the two counts are equal, the protractor is square. The same can be done for an east/west (northing) gridline using the 90 and 270 degree marks on the sides of the scale. You can use the same technique with the outer Mil scale, apply the same logic and use mils at the 0, 1600, 3200, and 4800 mil cardinal points.

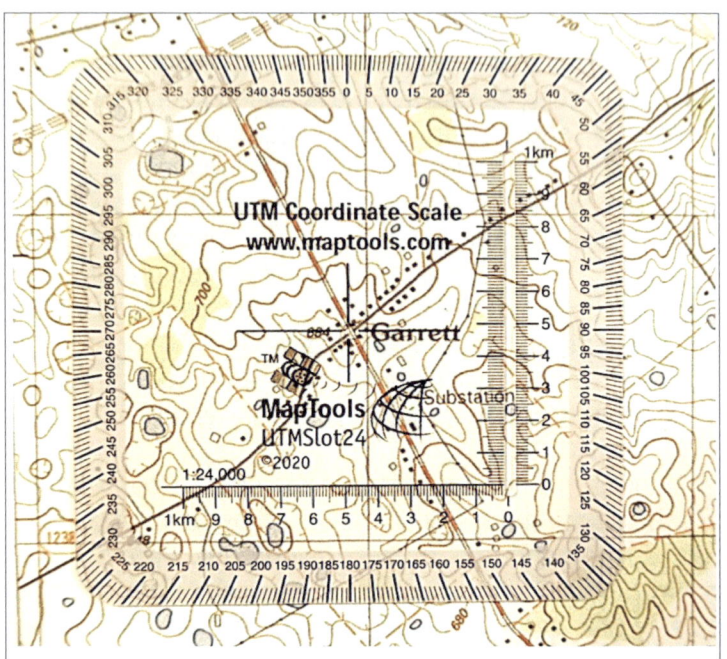

Example of a "squared up" protractor on a map; this is set up to measure a grid azimuth from the road intersection in the town of Garrett. Use a grid line that crosses your protractor to ensure it crosses the degree scale at directly opposite points. In the example, there is an easting that crosses at exactly 4 and 176 degrees. Both marks are 4 degrees east of the 0/180 degree north/south line, so you know it is parallel and perpendicular to the grid lines on the map (aka squared up).

Measure (Determine) a grid azimuth

During route planning your first leg might be a dead reckoning movement (requires a distance and direction). You need a direction (an azimuth) to use on your compass, in this case you would get that info from your map first. To get this grid azimuth you must first plot your points from your route planning.

When plotting navigation points make a small precise dot and then draw a larger circle around it. The circle helps you find the smaller dot without enlarging the actual point. If you already have a 6 or 8 digit grid for a nav point or waypoint plot it with your protractor using the procedure from Chapter 2 (pages 64-66).

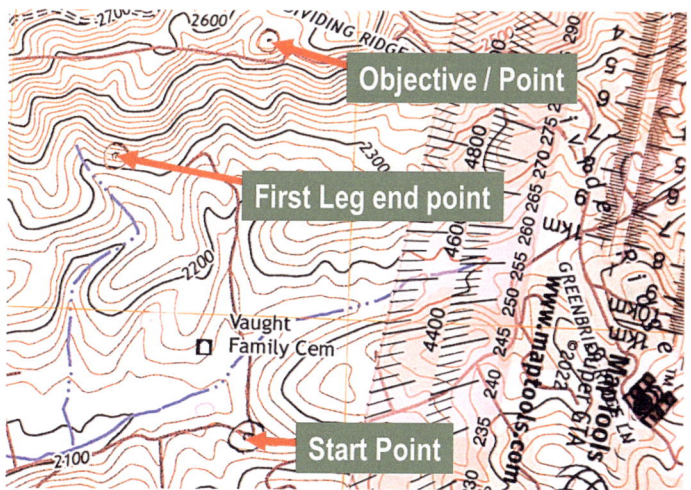

Mark your start and end point(s) on your map.

Next you will draw a line on your map with a straight edge from the start to the end point of your first leg. Draw the line through the objective (or point) to ensure it will extend past the degree scale of the protractor when you take the reading. Around 2k long for a 1:24 map and 4k for a 1:50 is adequate; no need to measure it, with experience it will be something you do naturally.

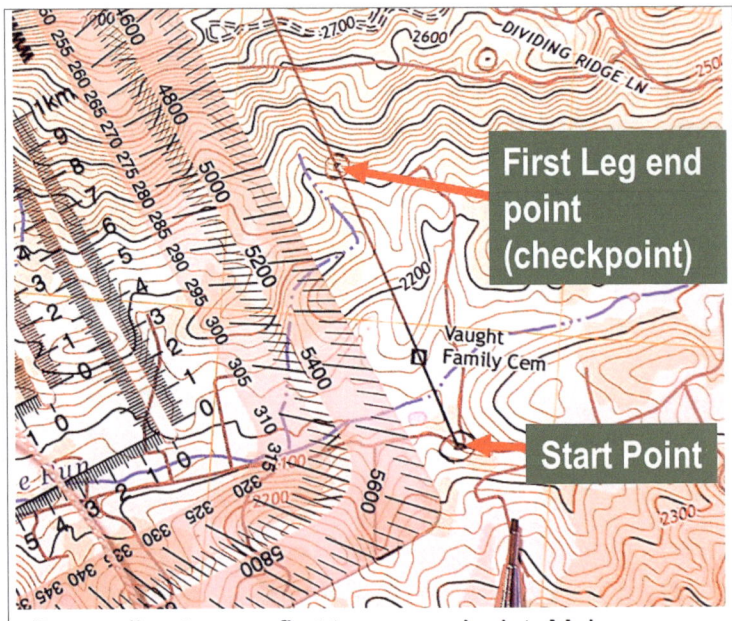

First Leg end point (checkpoint)

Start Point

Draw a line to your first turn or endpoint. Make sure you draw it long enough to reach the degree scale of the protractor so you can measure the angle, it is ok to draw it past the checkpoint. About 2km long is good to go for a full size GTA protractor.

Measure the Grid Azimuth. Place the center point (the index) of your protractor over the start point. Read the value of the angle on the degree scale; this is the grid azimuth (GAZ) of 331 degrees from the start point to checkpoint (see the next diagram). You can now use the GM angle to convert this grid azimuth (the one you just drew and measured) to a magnetic azimuth you can use on your compass.

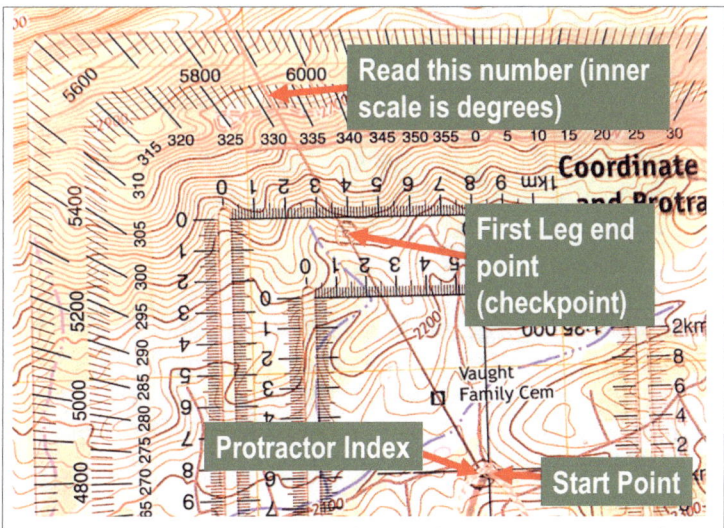

Measure the Grid Azimuth (GAZ) from start to end to the closest degree. The line we drew crosses between 331 and 332 degrees. I would rather miss to the left so I know my point should be to the right if I don't immediately see it - so Ill use 331. If this is a leg for a movement with no physical point a half degree over the 1k distance won't create much of a nav error. The point here is don't sweat minor interpretations like this.

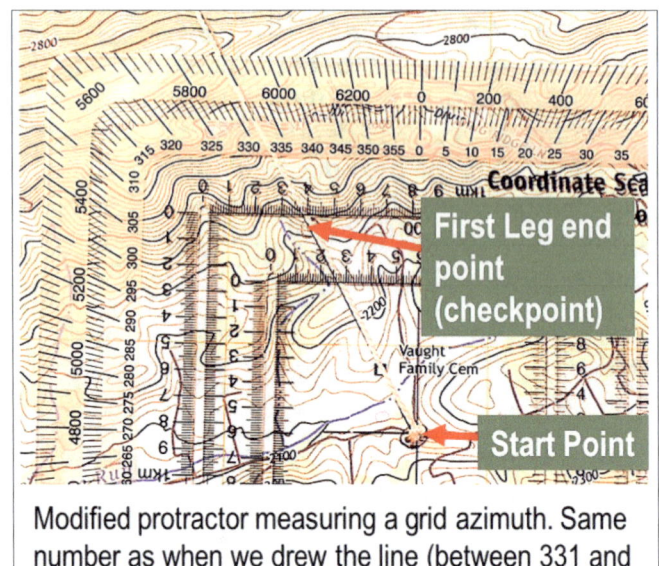

Modified protractor measuring a grid azimuth. Same number as when we drew the line (between 331 and 332 degrees GAZ).

Modified Protractor. You can modify your protractor by drilling a small (tiny) hole directly at the index line cross. Use a single strand of 550 gut (inner strand) that is 6-8 inches long. There is no standard for the length, use your judgment if you want to try this modification.

Thread it through the hole from the face (top) of the protractor. This will keep the knot from being between the map and the scale. It helps to melt the ends of the string before you thread it through the hole. Tie a knot in each end, now it can be used to measure direction without drawing on the map. You must be careful when using this technique, it is one more error inducing element to manage when you are taking readings - so it takes practice. I use mine to supplement the basic techniques, it isn't a substitute for sharp pencil map work.

Measure Distance. Now you have the MAZ for your route legs (derived from the GAZ you measured on your map and applied the declination math). Now you need the straight line distance for each of the legs.

If the distance is less than 1km (1,000 meters) you can simply use the scale on your protractor. Be aware your may have to flip multi scale protractors 180° to use the correct scale.

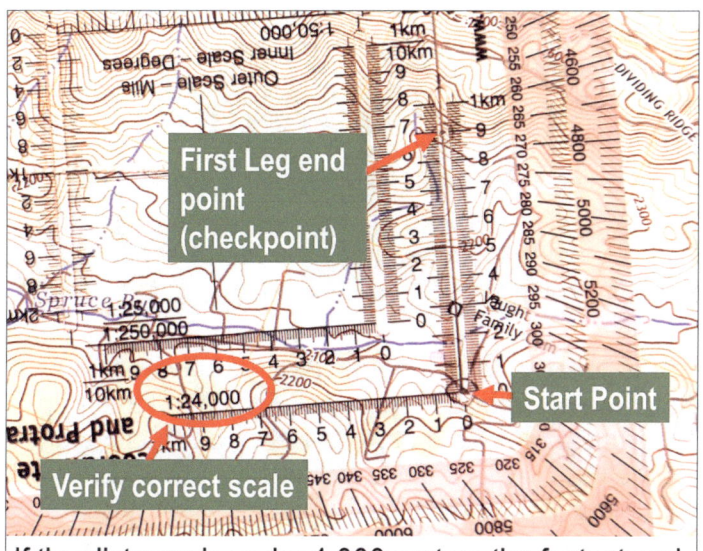

If the distance is under 1,000 meters the fastest and most accurate distance measuring tool is the protractor scale. Just by chance this first leg is exactly 900 meters.

Measuring distance greater than 1,000 meters. To measure longer distances you will use the map graphic bar scale in the marginal data. A graphic scale is a ruler printed on the map and is used to convert distances on the map to actual ground distances. The scale is divided into two parts, to the right of the zero is the primary scale marked in full units of measure. The extension scale is to the left of the zero and is divided into tenths. Topo maps can have three or more graphic scale bars, be sure to use the correct one to measure your navigation legs (meters/km).

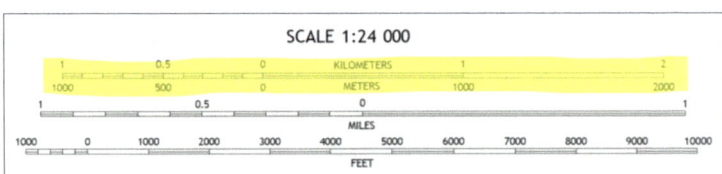

The highlighted KM/M bar on the scale is the one you must use. Unlike the declination diagram this is printed to scale so you can physically use it to measure distances on your map.

To determine straight-line distance between two point, lay a straight-edged piece of paper on the map so that the edge of the paper touches both points and extends past them. Make a tick mark on the edge of the paper at each point.

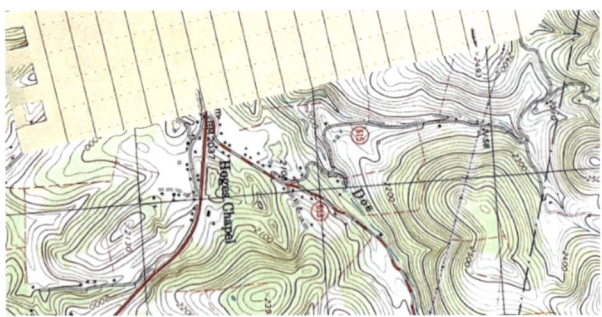

Measuring the straight line distance from BM 2037 to BM 2493

To convert the map distance to ground distance, align the right tick mark with a printed number in the primary scale so that the left tick mark is on the extension scale. The right tick mark is aligned with the 1,000-meter mark in the primary scale, so you know the distance is at least 1,000 meters. To determine the distance to the nearest 100 meters, look at the extension scale to the left of zero. The extension scale increases to the left so always read it right to left. It is divided into hundred meter increments.

To determine the distance from the zero to tick mark you can mentally divide the distance inside the squares into tenths. A more accurate method is to measure the "last" 1,000 meters of your line with your protractor. This is the preferred method since you don't have to interpolate the measurement blocks as you do on the map scale.

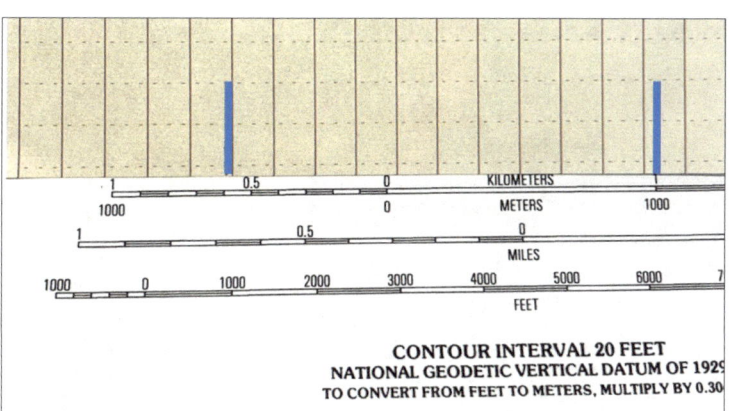

Put the right hand mark on the farthest mark from 0 that still keeps the left hand tick mark on the extension scale (left hand scale)

139

Measure distance along a road

To measure distance along a road, stream, or other curved line, use the straight edge of a piece of paper. Place a tick mark on the paper and map at the beginning point of the feature you need measured. Align the edge of the paper along a straight portion and make a tick mark on both map and paper when the edge of the paper leaves the straight portion of the line being measured. Keeping both tick marks together (on paper and map), place the point of the pencil close to the edge of the paper on the tick mark to hold it in place and pivot the paper until another straight portion of the curved line is aligned with the edge of the paper. Continue in this manner until the measurement is completed. Ensure you stay on the same side of the linear feature when you use this technique. Now move the paper to the graphic bar scale on your map and measure between are the start and end tick marks to determine the ground distance.

Measuring a section of road distance along RTE 730. You are marking on the map and the paper together, rotate to keep it against the south side of the road. Rotate to the curves, realign the last two tick marks (map and paper) and keep on marking and rotating. This is the result.

SHOOTING AN AZIMUTH

After determining your location and planning your route you now know the distance and azimuth of your first leg so it is time to take your first azimuth and step off. Shooting an azimuth is the task of aiming your compass at a distant object or aligning the compass with a specific degree reading. Some will refer to this as taking a bearing, but for consistency in the US we refer to it as an azimuth.

Centerhold Technique (Lensatic). Open the compass to its fullest so that the cover forms a straightedge with the base. Move the lens (rear sight) to the rearmost position, allowing the dial to float freely. Next, place your thumb through the thumb loop, form a steady base with your third and fourth fingers, and extend your index finger along the side of the compass.

Place the thumb of the other hand between the lens (rear sight) and the bezel ring; extend the index finger along the remaining side of the compass, and the remaining fingers around the fingers of the other hand. Pull your elbows firmly into your sides; this will place the compass between your chin and your belt. To measure an azimuth, simply turn your entire body toward the object, pointing the compass cover directly at the object. Once you are pointing at the object, look down and read the azimuth (degrees are RED, think "Read RED") under the fixed black index line.

Lensatic compass centerhold technique

The centerhold method is faster and easier than compass to cheek and can be used under all conditions of visibility. Remember to keep your carbine and other metal objects away from the compass to prevent magnetic interference. Mind your magnetic hydration tubing holders, they can be a hidden culprit.

Metal objects and electrical sources can throw off a compass needle. The following separation distances are suggested to mitigate interference:

High-tension power lines: 55 meters.

Vehicles: 10 meters.

Telephone wires and barbed wire fences: 10 meters.

Carbine/Rifle 1/2 meter (probably the most important for us).

Centerhold (Baseplate compass). For a baseplate centerhold the technique is similar, the key being a level, stable platform when taking a reading. Rotate the bezel on a baseplate compass to align the dial to the desired bearing or direction. Now hold the compass level and rotate your body until the red end of the needle aligns with the orienting arrow (the doghouse or shed looking arrow). The compass direction arrow (the small arrow printed on the compass frame that does not move) is now oriented toward the direction you want to walk if you keep the needle pointed at the orienting arrow. Some will refer to placing the needle in the orienting arrow as "red in the shed." Find a distant landmark the direction arrow is pointing, put your compass away and walk to the landmark. Verify your direction with the compass as you repeat the process.

Baseplate compass centerhold technique is the same as with the lensatic. One caution with using a baseplate, double check the bezel and indexed azimuth each time. It is easy to bump or turn the bezel on some baseplate compasses.

Compass-to-Cheek Method (Lensatic). Fold the cover of the compass containing the sighting wire to a vertical position; then fold the rear sight slightly forward. Look through the rear-sight slot and align the front-sight hairline with the desired object in the distance. Then glance down at the dial through the eye lens to read the azimuth (read RED). The compass-to-cheek technique is used almost exclusively for sighting, and it is the best technique for this purpose. Baseplate compasses with mirror sighting systems can also be used in a similar manner.

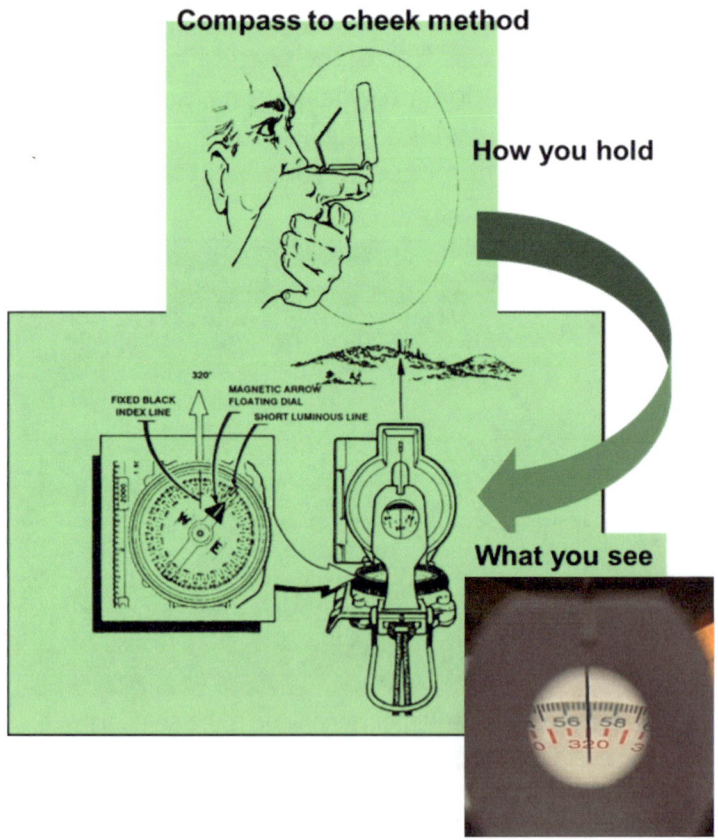

Compass to cheek method

How you hold

What you see

144

PRESETTING A COMPASS AND FOLLOWING AN AZIMUTH

Compass models will vary slightly in the details of their use, but the principles are similar.

Setting up an Azimuth for Daytime (or with a light source at night):

Hold the lensatic compass level in and rotate it until the azimuth falls under the fixed black index line. Now turn the bezel ring until the luminous line is aligned with the north-seeking arrow to preset the azimuth.

To follow an azimuth, use the centerhold technique and turn your body until the north-seeking arrow is aligned with the luminous line and verify the actual degree with the index line. Find a steering mark in line with the front cover's sighting wire, which is aligned with the fixed black index line that contains the desired azimuth. When you stop to check your azimuth again, use the actual number under the index line (for daytime). We do this since the luminous mark has 3 degrees in every click. You always want to remove as many errors as possible, this is a simple and trusted technique vs only using the luminous mark.

For a baseplate compass the technique is a bit simpler, but not as precise. Let the compass settle, keep the needle inside orienting arrow ("red in the shed") and walk in the direction of the baseplate / direction of travel arrow.

Setting an Azimuth at Night

When using your compass at night set the initial azimuth while light is still available (if possible). With the initial azimuth as a base, any other azimuth that is a multiple of three can be established by using the clicking feature of the bezel ring (3 degrees per click). The azimuth may be set on a baseplate compass that has luminous markings by using the illuminated window to read and set the degrees under the index line. If your baseplate model only has needle/orienting arrow and direction arrow illumination, you will have to set the azimuth using a low power red lens flashlight.

For a lensatic compass rotate the bezel ring until the luminous line is over the fixed black index line. (this click technique is used when you cannot read and align the degree to the index line and just align the marks as we do for daytime). On a tritium compass it is easy to see the degree numbers with the naked eye at night and just align the marks. To be honest this click and count technique is a bit antiquated and not that useful most times. But here is how it works:

Divide the desired azimuth by three for the number of clicks you need to rotate the ring. If the desired azimuth is smaller than 180°, the number of clicks on the bezel ring should be counted in a counterclockwise direction. For example, the desired azimuth is 51°. Desired azimuth is 51° ÷ 3 = 17 counterclockwise. If the desired azimuth is larger than 180°, subtract the number of degrees from 360° and divide by 3 to obtain the number of clicks. Count them in a clockwise direction. For example, the desired azimuth is 330°; 360°-330° = 30 ÷3 = 10 clicks clockwise.

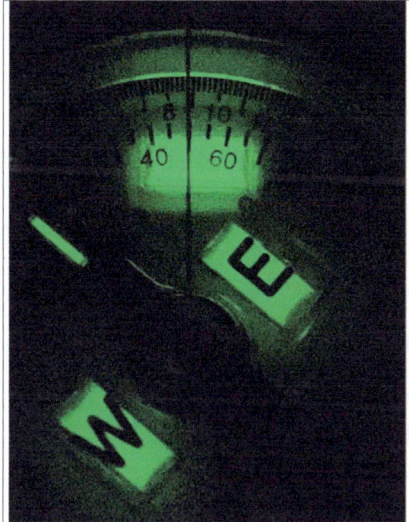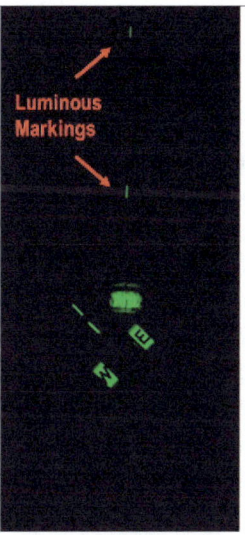

Lensatic set for 50 degrees magnetic at night. The photo exposure is not enhanced, this about what you can see on a tritium Cammenga without using an external light source. You can always read the degrees at night. The "3 degree per click" method to set a lensatic at night is unnecessary under normal conditions with decent eyesight.

With either type of compass preset as described above, assume a centerhold technique and rotate your body until the north-seeking arrow is aligned with the luminous line on the bezel (lensatic) or the needle is between the luminous markings of the orienting arrow (baseplate). Find a steering mark along the direction of the front luminous dots and move out. Re-shoot the azimuth as often as needed.

Sometimes the desired azimuth is not exactly divisible by three, causing an option of rounding up or rounding down (for the lensatic compass) which will induce a small but unavoidable error into the process. Don't worry though, you can be confident this will not throw your night navigation off.

Plot an azimuth from your compass on a map. You may be provided with magnetic azimuth(s) for a mission, or you may see a distant you want to plot on your map. In the second case you need to shoot an accurate azimuth to the object with your compass. Convert the magnetic azimuth from your compass reading to a grid azimuth using the GM angle and the LARS rule. Now you can plot the GAZ on your map.

Place the protractor on the map with the index mark at the center of mass of the known point (your location / where you shot the azimuth from). Square your protractor up so it is parallel to the map grids by using the technique we discussed earlier.

Make a tick mark at the azimuth along the degree scale of your protractor. On GTAs and some other models the mil scale is the outer ring so it can be hard to mark degrees. This is when the modified protractor with the 550 strand we talked about comes in handy.

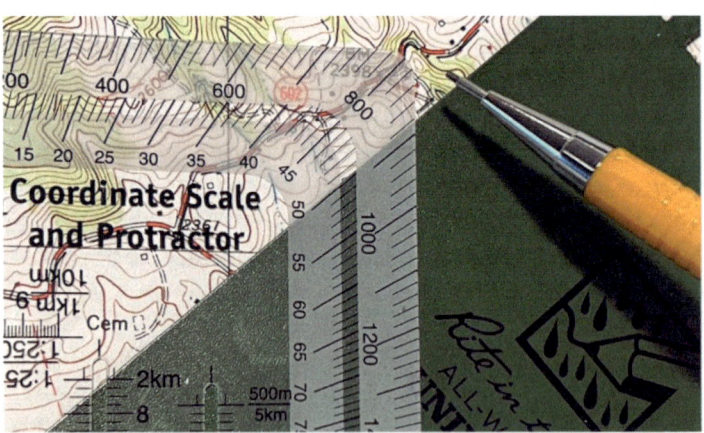

With the protractor index (the dead center cross) on your start or origin point square it up and mark the degree with a tick mark (pencil or superfine alcohol marker depending on waterproofing). If you have an outer mil ring using a piece of paper or straight edge will help you mark the degree accurately.

Remove the protractor and draw a line connecting the known point (your location in this case) through the tick mark. This is the grid direction line (grid azimuth).

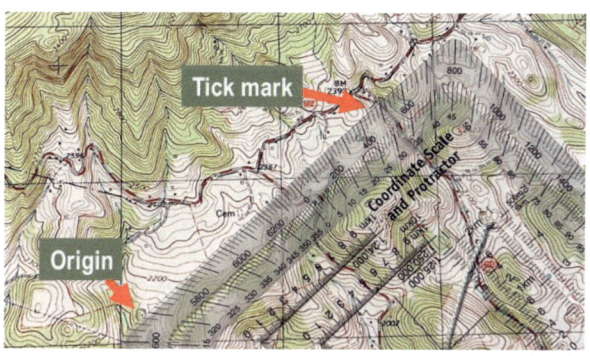

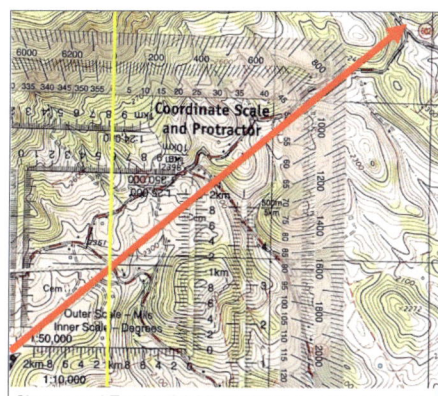

Verify your grid azimuth is correct (measure twice) to make sure you plotted it correctly. Place the protractor index where the grid azimuth line you drew cuts across an Easting or a Northing. Align the index (the center) of the protractor on the intersection of the azimuth line with the grid line. You can now easily (and precisely) determine your initial azimuth reading is correct.

Choose and Easting (highlighted with a yellow line in the photo) or Northing along the GAZ you drew. Align the protractor index along the grid line to check your work. This technique is a fast way to double check a GAZ on a map. Measure twice, walk once.

PACE COUNT

Determining Your Pace Count. We talked through the mechanics of finding a direction both on the map and with a compass as well as measuring map distance. We need a method to track the distance on the actual ground as we move. A way to measure this ground distance is the pace count. A pace for land navigation purposes is equal to two natural steps, counted on every other foot. Always step off each time with your left foot and count only the left foot strikes. Your daytime unloaded flat terrain 100 meter pace count is probably going to fall somewhere between the low 60s and low 70s.

To determine this, you must walk an accurately measured course (GPS or rangefinder) and count the number of paces you take. A pace course can be as short as 100 meters or as long as 600 meters.

The pace course(s), regardless of length, must be on varied terrain similar to what you will be walking over. It does you no good to walk a day course on flat terrain without your gear and then try to use that pace count on hilly terrain at night while carrying a heavy ruck. Day, night, load variance, terrain type, soil type (sand will significantly add paces), weather, and fatigue level will all influence your pace count. Experiment during your training to determine what works for you and what is the most accurate. Use and modify the Pace Count Record in Appendix E to keep track and help you memorize yours.

Keeping Pace Count. Maintaining count during a movement is critical, do not rely on your memory (was that 2 or 300 meters?). Losing count is a real thing under stress, fatigue, and mission focus. Using pace beads, moving pebbles from one pocket to another, or simply making tick marks in your notebook are all ways to keep track. Pace beads are hands down the most efficient and the least distracting method of keeping count. Legs are usually only a few hundred meters, but typical pace beads provide kilometer counts as well.

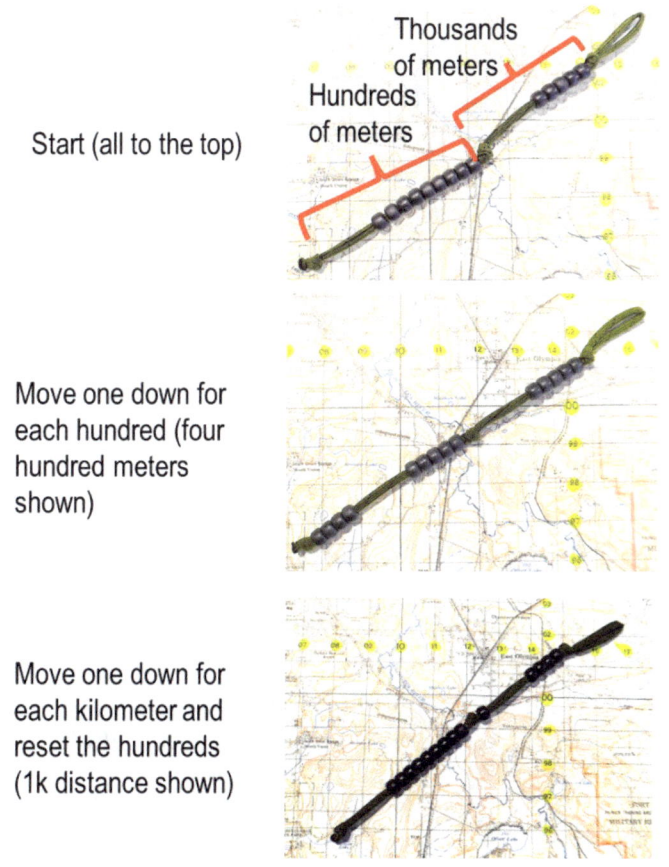

Thousands
of meters
Hundreds
of meters

Start (all to the top)

Move one down for each hundred (four hundred meters shown)

Move one down for each kilometer and reset the hundreds (1k distance shown)

Moving by Dead Reckoning. Dead reckoning is following an azimuth for a specified distance. You will use a protractor to determine the direction and distance from one point to another on your map. Once you have converted the grid to a magnetic azimuth using the GM angle/declination diagram you will use your compass and pace count to track your direction and distance to the objective.

Most dead reckoned movements do not consist of single straight-line distances due to the mission factors we discussed earlier. Another reason most dead reckoning movements are not single straight-line distances is because compasses and pace-counts are imprecise measures. The farther you travel by dead reckoning without confirming your position in relation to the terrain and other features the more errors you will accumulate. You must confirm (and correct) your estimated position whenever you encounter a recognizable feature on the map.

Dead reckoning is simple to learn and easy to teach to your team or family. It is an accurate way of moving from one point to another over short distances, even with few external cues are present to guide the movement (at night for example).

The downside is dead reckoning is time-consuming and demands constant attention to the compass. Every fold in the ground and detours as small as a single tree or boulder also complicate the measurement of distance.

During daylight, across open country and along a specified magnetic azimuth, never walk along with the compass in the open position in front of you. The compass will not settle or provide you with an accurate reading. This is a rookie mistake we see often; students bumbling along staring into their compass for answers when they should be looking at the terrain (and situation) around them. Begin at the start point, face the compass in the proper direction, then sight in on a landmark that is located on the correct azimuth. Close the compass and move out to the landmark. Repeat the process as many times as necessary to complete the straight-line segment of the leg.

These landmarks are called *steering marks*, and their selection is crucial to success in dead reckoning. Steering marks are selected as you move, they are usually on or near the highest points that you can see along the azimuth line that you are following. They may be uniquely shaped trees, rocks, hilltops, posts, towers, and buildings—anything that can be easily identified and kept in sight. If you do not see a good steering mark to the front, you might use a back azimuth to some feature behind you until a good steering mark appears out in front.

A steering mark should always be visible as you move toward it...and it should be stationary. Do not use objects that can move. I once had a student who shot an azimuth to a cow across a large open field and tried using it as a steering mark; never a good technique.

Choosing a prominent steering mark in the distance like this hill is great until you walk into the vegetation and can no longer see it. Steering marks will change often when dead reckoning, you may be shooting azimuths a few meters at a time when your line of sight is limited.

In areas of dense vegetation, terrain with little relief, night, or in fog, your steering marks must be close together. You may be shooting 5 meter azimuths in dense forest or jungle vegetation. These short, numerous steering marks introduce more chances for error. Steering marks at night must have even more unique shapes than those selected during daylight, even if using NVGs. Instead of seeing shapes, you begin to see only the general outlines that may appear to change as you move and see the objects from slightly different angles.

You may have to send a teammate out in front of you to create your own steering mark (open desert or plains, night, fog etc). To use this leap frog or bounding technique your teammate should be as far out as possible (METT-TC dependent). Hand and arm signals or a radio may be used in placing him on the correct azimuth (use caution). Move forward to their position and repeat the process until some steering marks can be identified or until you reach your objective.

OBSTACLES

When an obstacle forces you to leave your original line of march and take up a parallel one, always return to the original line as soon as the terrain or situation permits. To turn add or subtract from your current azimuth direction. If you are in dense, tight terrain and navigation is difficult use 90 degree turns so you can keep track of your pace count as you bypass. If you can terrain associate and "re-find" a new start point along the leg or route it is less critical.

Bypassing an Obstacle. To bypass threats, danger areas, or obstacles and still stay oriented you can detour around the obstacle by moving at right angles for specified distances (box method). Add and subtract 90 from whatever your travel azimuth is to get your turns.

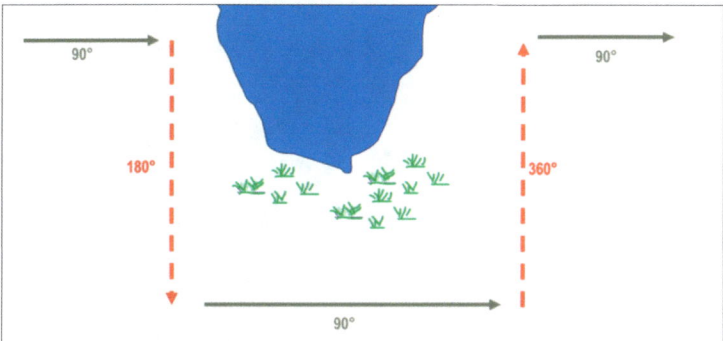

Detour bypass or box method. Total pace count does not include the red legs, but the red pace must match each other to stay on track for the turns to get you back on the original route.

When using the box method be certain that only paces taken toward the end of the leg or objective are counted as part of your forward progress. They should not be confused with the local pacing to execute the bypass (the two vertical legs in the previous diagram) that takes place perpendicular to the route.

We are talking about dead reckoning in this section, but you should apply the best combination of methods for the situation. If you can bypass by handrailing an obstacle or danger area that may be the better option. Maybe you can find a steering mark on the far side and then make your way to it by an indirect route. Whatever works - as long as you can pick your route back up and account for the straight line distance (not your bypass distance). Make your skills all work together to give yourself options.

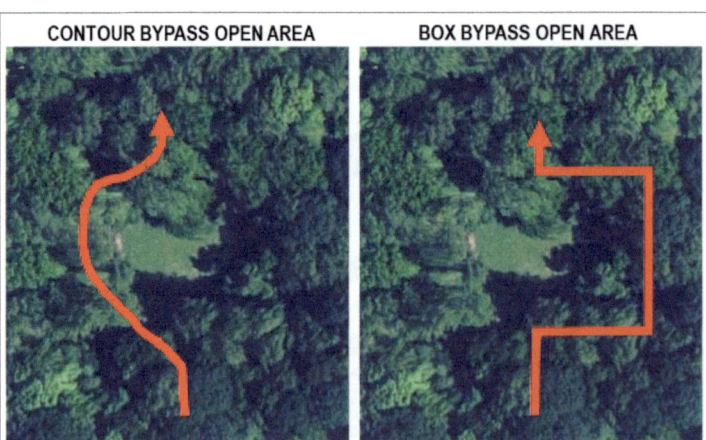

Two methods of bypassing an obstacle (or open danger area). Use the contour bypass (handrail) when you are confident of your terrain association abilities and there are good landmarks. You must be able to pick your route back up on the far side. For pure dead reckoning legs or when the terrain /visibility does not support terrain association the detour bypass (box) is the better method.

ALONG THE ROUTE

As you traverse your route you will be presented with navigational problems to solve. It goes back to Chapter 2 and the first step in the navigation process "know where you are." Sometimes you will need to verify that you are still on your route, or you may need to determine the grid location of a distant object. There are two manual methods we can use with a map and compass to do this. We can use a procedure known as an *intersection* to accurately plot a distant unknown / unmarked point (object or target) on the map. We can also use a similar type of manual technique called a *resection* to find our own location.

Land navigation is about problem solving. Be prepared to make adjustments to your route and have the skills and confidence to self-correct when things do not go according to plan.

INTERSECTION

An **intersection is used to determine the grid of an unknown stationary point** by shooting an azimuth from at least two different known positions on the ground. This is used to locate the grid coordinate of distant or inaccessible points or objects such as enemy targets and danger areas.

Using the map and compass method you will orient the map using the compass and mark your position. Determine the magnetic azimuth to the unknown position using your compass.

Convert the magnetic azimuth to grid azimuth and draw a line on the map from your position along this grid azimuth. This is the same process we used to plot an azimuth on a map (see page 146).

Move to a second known point and repeat the steps so you have a second line (grid azimuth) drawn on your map. The two GAZs will cross each other, the location of the unknown target is where the lines intersect. Now you can read the 8 digit grid of the intersection with your protractor. You can also measure the distance on your map to the target or object.

A third azimuth from a third known position will make this even more accurate. This technique works well if you have multiple observers at known points reporting the data.

INTERSECTION

Things you don't know:
- The grid location of the target
- The distance to the target

Things you must know:
- Your grid location on the map
- The azimuth to the target

(Know these from two separate locations)

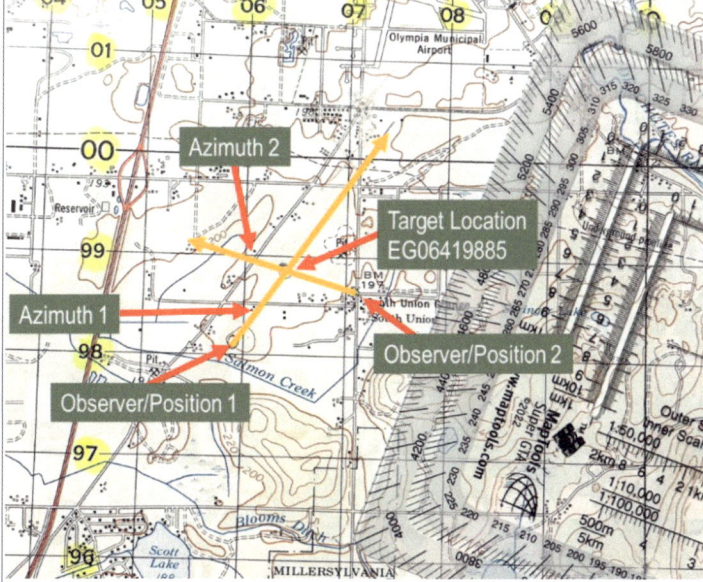

An intersection is two azimuths taken from two locations of the same target. Once both azimuths are plotted on the map the intersection of the two GAZs will give you the target's grid location.

POLAR COORDINATES

A polar plot is a method of **locating or plotting an unknown stationary position from a known point by giving a direction and a distance** (called polar coordinates). It is like the intersection technique but removes the requirement to have data from a second location and replaces this requirement with a distance to the target or object. You must have your own location plotted on the map, the azimuth to the object, and the distance in meters to the object to come up with polar coordinates. This is used if the target is not near an identifiable feature on your map (that you could use to figure a grid from that data alone).

POLAR PLOT

Things you don't know:
- **The grid location of the target**

Things you must know:
- **Your grid location on the map**
- **The azimuth to the target**
- **The range (distance) to the target**

POLAR COORDINATES

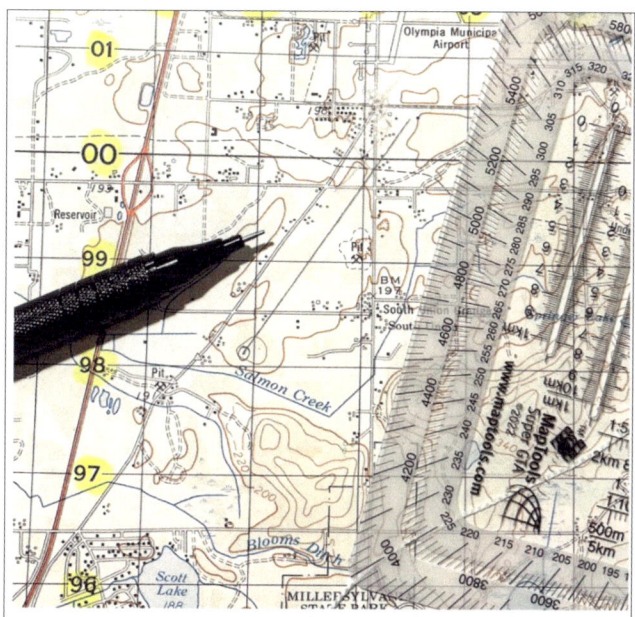

The observer's known position is EG 05909808 and the target is at range of 890 meters (by rangefinder) at 20 degrees MAZ (35 degrees GAZ). A polar coordinate lets you know the target is along this line (35 GAZ) at 890 meters.

The direction and a distance along that direction line is a polar coordinate; it consists of your present known location on the map, an azimuth (grid or magnetic) and the distance to the target (in meters).

ESTIMATE RANGE / RANGE DETERMINATION

Having a way to measure or estimate distances to objects or targets is a valuable tool for navigation. As you saw with the polar plot example there may be times when you need the range to a distant feature or target is not marked on your map. The best tool for measuring this is a laser rangefinder. Having a small rangefinder in your kit could be worth the slight weight and volume carry penalty. If you do not have access to a working rangefinder there are some manual means to estimate range - but they are *estimates* at best. For precision navigation purposes having a rangefinder is highly recommended.

A small handheld rangefinder can be a valuable tool when navigating. No estimating required; you will get the exact distance for an object. (DoD Photo)

Football-field Method. The length of a football field is 100 yards instead of meters, but it is a unit of measure many of you are familiar with. Estimate the number of football fields between you and the target for ranges up to 500 meters. Multiply the number of fields by 100 to estimate the range in meters. As you can see this leaves a good bit of room for error and should be used with this in mind.

Other methods include the *recognition method*, *flash-to-bang*, and the *reticle (mil relationship / wrm formula method)*. We won't go into detail on these in here, just know there are several techniques you can learn to help you estimate range.

Using a monocular, binoculars, or rifle optics with reticles will assist with range estimation but none of these are a substitute for a good rangefinder. (DoD Photo)

RESECTION

A resection is the method to determine your grid location on a map by plotting the grid azimuth to at least two well-defined locations that you can pinpoint on the map and see on the ground. For greater accuracy you can use three distant features.

When using the map and compass method orient the map using the compass. Identify two (or three) known distant locations on the ground and mark them on the map.

Shoot an azimuth to one of the known positions from your location and convert the magnetic azimuth to a grid azimuth.

Convert the grid azimuth to a back grid azimuth. This is adding or subtracting 180 (see page 120). Using a protractor, draw a line for the back grid azimuth on the map from the known position back toward your unknown position.

Repeat these steps for a second position (and a third position, if needed). The intersection of the lines is your location. Now you can determine a precise grid coordinate of your location with your protractor.

RESECTION

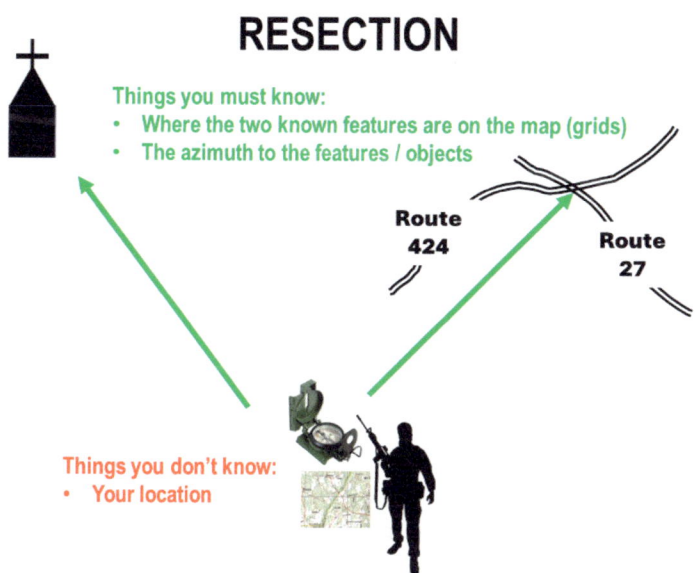

Things you must know:
- Where the two known features are on the map (grids)
- The azimuth to the features / objects

Route
424

Route
27

Things you don't know:
- Your location

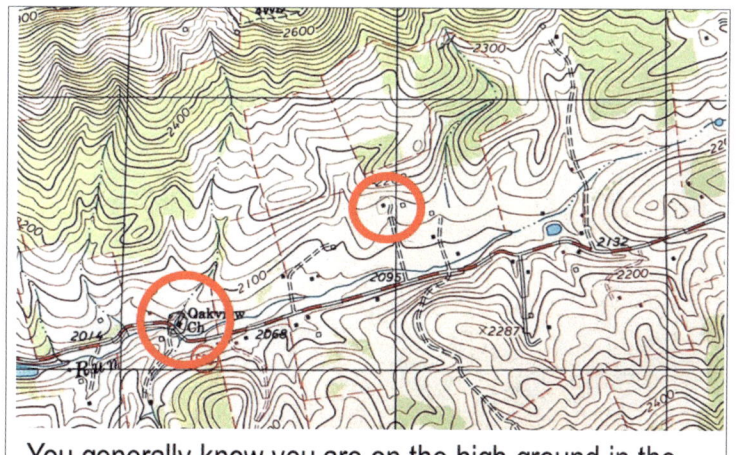

You generally know you are on the high ground in the
4025 grid square. You can see Oakview Church at 228
MAZ down by the hardball road and a building at the
end of the unimproved driveway at 106 MAZ.

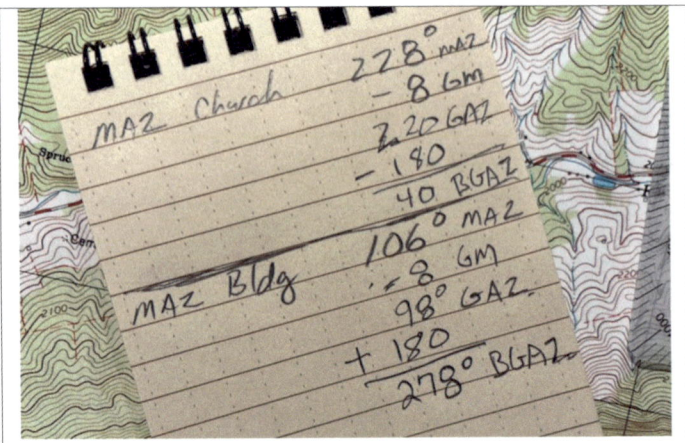

Do the math to get the grid azimuth (GAZ) and then the back grid azimuth (BGAZ). Don't be shy about writing everything down, it will help you catch any errors.

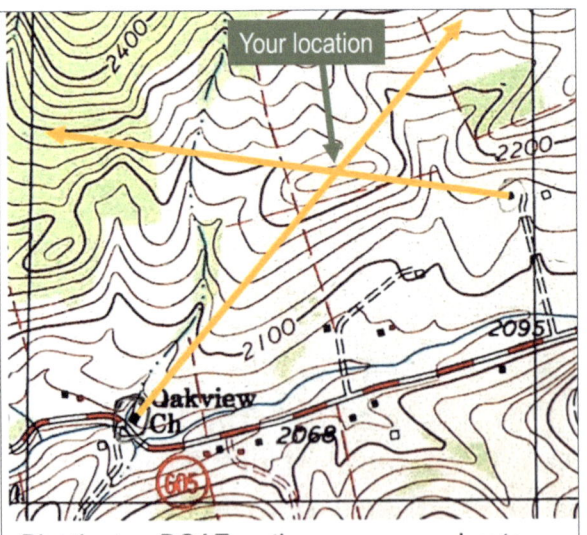

Plot the two BGAZ on the map, remember to start from the two known points. Your location is where the two BGAZs cross. Now you can use your protractor to get your 8 digit grid.

MODIFIED RESECTION

Modified resection is a method of locating your position on the map when you are on a linear feature such as a road or stream. The process is similar to the resection but you only need one known point in the distance.

Orient your map and find a distant point you can identify on the ground and on the map. Shoot an azimuth from your location to the distant known point.

Convert the magnetic azimuth to a grid azimuth then c*onvert the grid azimuth to a back azimuth.* Using a protractor, draw a line for the back azimuth on the map from the known position back toward your unknown position.

Your location is where the line crosses the linear feature. Determine the grid coordinates with your protractor.

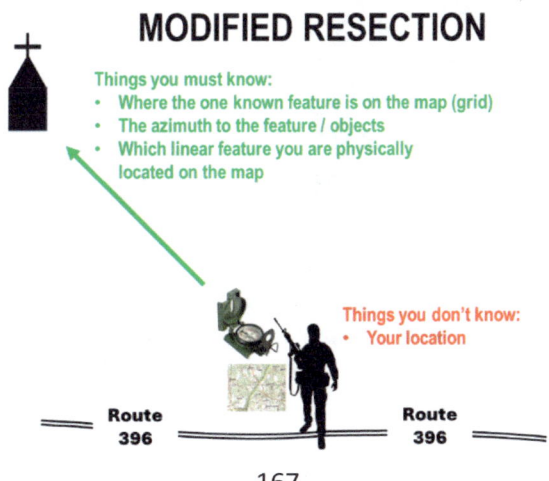

MODIFIED RESECTION

Things you must know:
- Where the one known feature is on the map (grid)
- The azimuth to the feature / objects
- Which linear feature you are physically located on the map

Things you don't know:
- Your location

Route 396

Route 396

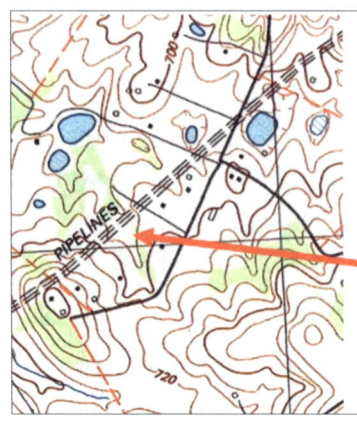

For a modified resection you need to know which linear feature you are standing on and the general location on the map.

This pipeline is a linear feature that is easily identifiable on the ground and on the map.

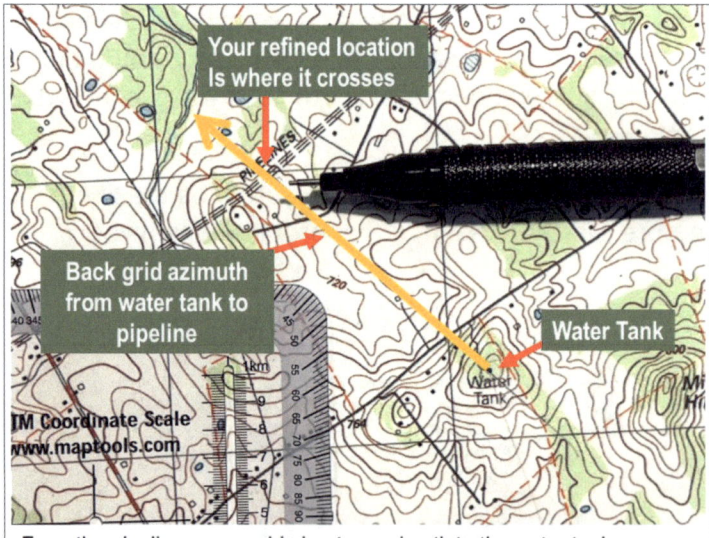

From the pipeline you would shoot an azimuth to the water tank you see on the ground (and is also a marked feature on the map). Convert to a grid azimuth (GAZ), then convert to a back grid azimuth (BGAZ) and plot it on the map *from the water tank back to your location* on the pipeline. Your exact location is where the line crosses the pipeline.

TERRAIN ASSOCIATION

The technique of moving by terrain association is more forgiving of mistakes and faster than dead reckoning. Navigation errors made using terrain association are easily corrected since you are comparing what you expected to see from the map to what you see on the ground. Periodic position-fixing through either plotted or estimated resection will also make it possible to correct your movements or report accurate grids to your team.

Orient the Map. The first step when using terrain association is orienting the map. A map is oriented when it is in a horizontal position with its north and south corresponding to the north and south on the ground. You may hear some say "orientate" but this is incorrect, you don't "orientate" a map.

Orient the Map Using a Compass. When orienting a map with a compass, remember that the compass measures magnetic azimuths and your map needs to be oriented using the grid azimuth to align with the actual terrain. Use your judgment on applying the GM angle to account for the declination; a rough orientation of your map is usually adequate, especially if the GM angle is not extreme and you have prominent terrain features. 99 times out of 100 you don't need to apply the GM angle to orient a map for practical way. Go with magnetic north on your compass to get close and then use the terrain to mesh the map info with the actual ground in front of you.

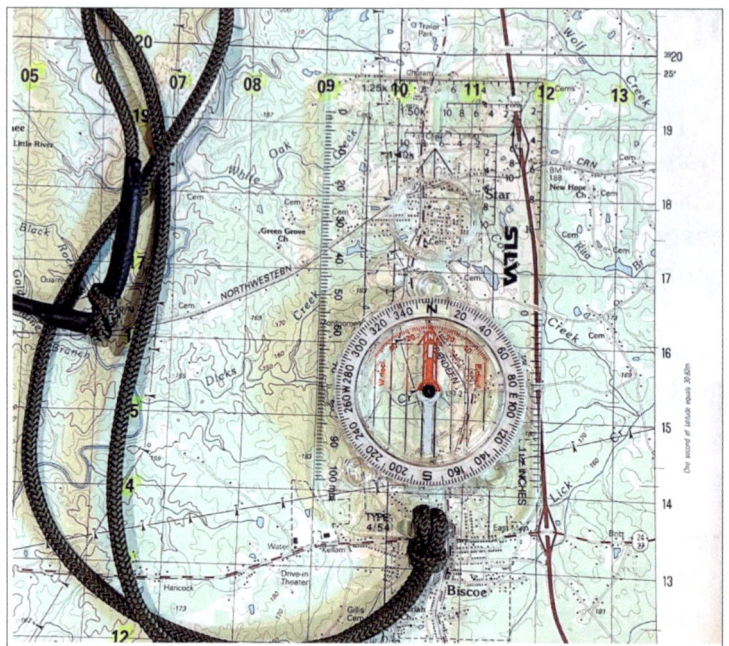

The current declination is 8 W for this area, but I'm comfortable with just using a magnetic azimuth to initially orient the map to the terrain. There is an abundance of features on the ground I can match up to orient it by terrain. Do not take this "good enough" shortcut when plotting azimuths and grids, the precise methods are the only way for those.

With the map horizontal on the ground or on a non metal surface, take the straightedge on the left side of the compass and place it alongside a north-south grid line with the compass pointing toward the top of the map. Now rotate the map and compass together until the magnetic arrow is at the index line on the compass (North). Your map is now generally oriented and that is usually adequate to allow you to interpret the map and terrain together in your mind.

If you have a desire to refine the orientation just do the declination math and calculate a magnetic azimuth for 0° grid north. Rotate your compass to the grid azimuth value (compass type dependent) while keeping the map grid line/compass alignment.

This is a good time to bring up a practice that some experienced navigators use. This may rub some navigation experts the wrong way (especially regarding using a baseplate compass on a map) but unless you lost your primary and alternate protractor / scales, I recommend never using your compass to measure or mark azimuths directly on the map. This can add errors to the process. Use the procedures from this manual to mark grid locations, draw azimuths, and take map readings with your protractor. Do the math to apply the declination with your protractor, do not try to physically align a map and then use the compass to draw the associated azimuth. It isn't "wrong", but it is a technique shortcut that carries too many error risks.

Orient Map Using Terrain Association. This is the primary way to use a map for terrain association. Hold the map so it is oriented (facing) the same way you are looking at the terrain (usually in the direction of movement) and match the features you see with the map sheet. Orient the map like this when you need to make many quick references while moving cross country. Using this method requires an understanding of terrain features and how they appear on the ground. The academic learning of terrain features from Chapter 2 coupled with a lot of practical application and practice is the only way to build this skill.

Know where you are starting. Back to the first step in the navigation process, *know where you are*. You must know your initial location before starting any movement along a route. If you move out trying to terrain associate before establishing your initial location, every turn or leg is going to be incorrect and will continue to compound navigational errors. It is a fact of human nature you can make the map match terrain that is nowhere near correct. This is especially true for newer navigators and when you are under stress. You must get a locked in start location from the beginning things will only get worse. Determine your initial location by referring to the last known position, by grid coordinates and terrain association, or by locating and orienting your position on the map and ground by resection.

Match the terrain to the map by examining terrain features and their relationship / alignment to each other. The planimetric maps we discussed in Chapter 1 are a good example of how terrain association works. The zoo or amusement park map is pure terrain association (probably more like "feature" association but the concept is the same). You are looking at the map and your surroundings and matching them up or associating them. If "x" feature is over there, you expect to see "y" feature next to it because they are that way on your map. Open terrain with prominent features makes this an easy task; dense woods with limited sight lines or low feature desert terrain make this more difficult. This reinforces the need to use all the tools available, a mix of compass and map work is always the best option if the environment supports it.

Identifying and Locating Selected Features. Being able to identify and locate the selected features, both on the map and on the ground, is essential to using terrain association. Look for elevation changes to locate and identify features being used to guide the movement. Look for the steepness and shape of the slopes (see Chapter 2, page 87), the relative elevations of the various features, and the directional orientations in relation to your position and to the position of the other features you can see.

Use Handrails, Catching Features, and Navigational Attack Points. Instead of using point navigation (dead reckoning) compass work, you can use a general azimuth without the use of steering marks if you can handrail the feature to guide your route. METT-TC applies, if you are in a semi- or non-permissive environment your options will be narrowed.

Vegetation. Sometimes you can use the vegetation to match the clearings, orchards, and treelines to the map. This is not optimal as these may not be depicted correctly (or at all) and can change rapidly due to cutting, growth, and planting. Some important vegetation features were likely to be different when the map was made. Another important factor about vegetation is that it can change rapidly through forest fires, developmental clearing of land, or farming. When comparing these you must be familiar with the different symbols, such as vineyards, plantations, and orchards that appear on the legend. The age of the map is an important factor when comparing vegetation.

Hydrography. Inland bodies of water can help during terrain association. The shape and size of lakes in conjunction with the size and direction of flow of the rivers and streams are recognizable features that are easy to find on a map.

Using Man-made Features. Man-made features are an important factor during terrain association. The user must be familiar with the symbols shown in the legend representing those features. The direction of buildings, roads, bridges, high-tension lines, and so forth make the terrain inspection a lot easier; however, the age of the map must be considered because man-made features can fluctuate. The reverse can be true as well, older maps from USGS (the "historical" maps we talked about in Chapter 1) have greater detail of manmade features than the new series national maps.

Terrain walk / recon your area during different seasons. In areas where vegetation and water features change with the seasons, a detailed examination of the terrain should be made during each change. The same piece of land may not present the same characteristics during both summer and winter. For example, in the woodlands sight lines open up in winter and may reveal unmapped features that were hidden behind foliage in spring/summer. The spring rains in some parts of the desert can bring flash floods that will wash out wadis.

Train and practice in the field. Terrain association failures are a training and experience issue. You must be able to interpret the map and analyze the terrain around you; you must bring the map and the ground together in your mind and make it make sense. Recognition of terrain and other features, the ability to determine and estimate direction and distance, and knowing how to do quick position fixing are skills that are more difficult to learn compared to dead reckoning. The use of terrain, other natural features, and any man-made objects that appear both on the map and on the ground are something you must practice often. There is no other way to learn or retain this skill.

The only way to get good at terrain association is in the field. The knowledge and technical skills are just the beginning.

A simplified example of terrain association: Starting at BM 2130 (near 1) this is how I would terrain associate to reach my objective vic NB 37472940.

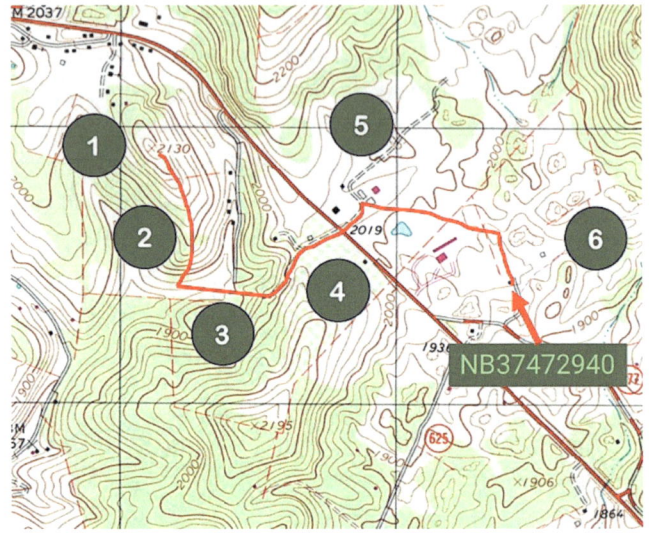

After planning my route on the map I will verify my location. The BM is marked but unverified; based on the map spot I know it is at the highest point on the hill. Since I am not using the benchmark to dead reckon from as long as I start at the hilltop I know I am good to go. I take a general azimuth of 170 to get me pointed in the direction of the ridge running north/south (2). I know I need to continue downhill while staying on the highest point along the ridge and then the lower spur line as it traces southwest. Once I reach the flat lowground (3) I will handrail the high ground to my south for about 300 meters where I will see the dead end of an unimproved trail to my left (north). I will follow the small spur for 150 meters along a 92 degree MAZ (using my pace count now) until I pick up another unimproved trail (west of number 4). There is a large ridgeline to my right (southeast) as I continue to make the curve along the trail. The trail will put me out on the west side of an orchard. I will continue to follow the trail NE until I cross the hardball road. I know when I am at the trail/hardball intersection I will have high ground behind me to the west and south and the terrain to the northeast becomes flat and open (number 5). 120 meters past the hardball intersection the unimproved road I am on makes a left hand curve near some buildings, I know that is where I need to leave the unimproved road and head back to the east toward the objective. I can see a single building (just west of number 6) that I had spotted on my map when planning the route. I will use it as a steering mark (no compass needed, I know where it is on the map and I know that is my last checkpoint before the objective). I will walk to the building keeping the two pieces of large high ground with the valley running north/south to my north. Once I reach my attack point (the building) I will take another general azimuth at 170 and I should see the objective building around 200 meters away. The hardball (just north of the grid marking box in the diagram) is a good backstop or catching feature. I know I have overshot the objective if I were to cross it.

This is a bit oversimplified, there are hundreds of these internal conversations and calculations when navigating. It is a constant stream of cross checks and verifications to confirm you are oriented. The example is a purely navigational one, if there were other factors in play (post disaster, hostiles etc) I would choose a different approach based on the METT-TC factors.

RECOGNIZE THE OBJECTIVE

The last step in the navigation process is recognizing the objective. This sounds simple, but in the woods or the desert it becomes not so clear at 0200 in the morning. You must set yourself up for success before you ever step off on the route. During route planning you must find identifiable features you can use to fine tune your final approach. An attack point near an objective will minimize the chance of navigation errors during your last short leg. A dead reckoning 50 meter leg from the attack point becomes an easy task, even at night in the rain. METT-TC always applies, a close in attack point may not be feasible if there is a hostile entity near or on the OBJ. The learning point is for you to build in every advantage you can.

Objectives and points are not always well defined. Give yourself every advantage before you get to the end of your route. Plan and use features along the way to build confidence in your location and apply all the skills you know to get to that final critical 50 meters.

COMMON (AND NOT SO COMMON) MISTAKES

Recognizing your objective at the end of a leg is the culmination of all the things we discussed earlier. Errors in navigation can take you just a few meters off the point or it can throw you off into a whole other grid square. We all make mistakes, but here are some common ones we have seen over the years that you should guard against.

Back azimuth. Watch your protractor placement and the direction you are measuring when you plot a grid azimuth. It is easy to go from the end to the start and get yourself 180 degrees wrong (literally). Always measure in the direction of travel unless you are intentionally calculating a back azimuth.

Lensatic lens lock down. This is a rookie mistake and some of you are shaking your head "no way" right now. I have seen several troops take a centerhold reading with their lensatic compass and then fold the lens assembly down to "lock in" their reading so the compass won't move as they navigate.

Rushing. Take your time when plotting and doing math. Don't be in a hurry to get started walking, even on timed courses. Work rapidly and deliberately, but take the extra time to double check your work. Have a mental checklist before you step off on a mission or on a nav leg.

GM Math errors. Math errors (reversing addition and subtraction) are common. Write the numbers down in your notebook when planning your routes and legs. No matter how simple the calculation, the time and effort are well spent to write it down. This allows you to see and review your work. Do the math twice to check your work. Label your work with GAZ, MAZ, BGAZ, and BMAZ ("B" is for back azimuth). Break it down and write it down, it will help you catch any errors.

Walking with compass open. We talked about this one earlier. Walking your route during the day while staring into your compass will induce navigation errors. Take your reading and move out to the steering mark.

Losing gear. Use lanyards on your baseplate and lensatic compasses and tie them down to your kit. Lanyards or dummy cords must still allow you to keep the item accessible and functional. Maps are easy to lose as well, if you stow your map in a cargo pocket make sure you secure the pocket to prevent losing your map.

Long legs. Newer navigators like to have a few or even just one long nav leg. Don't be afraid to add legs to break up the route and take advantage of the terrain. Build in as many opportunities to discover and correct errors along the route. The tradeoff for an extra 100 meters of walking may be worth doing if it takes you buy a recognizable feature. The small confidence checks along the way are nice to have.

Using the incorrect scale. This is easier to do than you might think, especially at night and when you are fatigued. Multiple scale protractors are excellent options, but it is easy to use the 1:25 scale on the 1:24 map. Navigation is attention to detail and checking your work, doing both will help prevent this type of error.

Reading Up then Right. Another common mistake for new navigators is to read grids up then right. This is a matter of getting reps in and always reading right then up.

Wrong datum in GPS. The default "just put in WGS84" is a bad habit many military GPS users get into. Check your map datum and use the one that aligns with your map version.

Do not assume all your maps and all your teammates are using the same datum. An easy error to make is being complacent about checking what datum is in your system.

CHAPTER 5
MOUNTED LAND NAV

This chapter briefly discusses some highlights of Mounted land nav. Mounted has the same components as dismounted land nav but with a few mental adjustments to account for the faster speed and limited routes and usable terrain for a vehicle. Moving by a typical car or truck, mountain bike, motorcycle, and even public transportation are all mounted movements. Movement can be administrative and routine or moving tactically as part of a mounted unit formation.

Mounted Land Navigation

The context for this short chapter is a set of abnormal conditions we might face; riots, post-disaster recon, or regional societal collapse / WROL. We felt like this is worth discussing in a focused land nav manual, especially since it can be overlooked in our community.

The principles for mounted land navigation are the same as dismounted land nav. The major difference between dismounted and mounted is the speed of travel and the limited mobility corridors and routes we can use with larger vehicles. We won't discuss water or air movements, only ground vehicle movement.

You will have the ability to cover long distances quickly, so you must develop the ability to estimate how far you have traveled and have a sense for how long routes will take. Using the odometer on the vehicle, time distance factors, and even plain old experience can assist with distance traveled. It is a package deal, you must combine and use all the skills you have learned. Terrain association, direction finding, map reading, and terrain analysis all come together.

Even mountain bike / bicycle movement changes the way you will navigate. The speeds and terrain considerations will require you to think faster and make rapid navigation decisions on the fly.

Things move fast for mounted land nav, especially when you are processing other tactical information while you move. You must shift your thinking from the 4 mph walking mindset to mentally stay ahead of a route in a vehicle. Break out of the interstate road trip mentality and keep this in context.

Mounted land nav is best done using GPS coupled with map and terrain orientation due to the speeds and required workload of a vehicle commander / navigator (that is you if you are in the front passenger seat or if you are on a single rider vehicle). Terrain orientation is the next best option with dead reckoning a distant last.

Mounted land nav is becoming a lost art. Get your team or family members used to navigating without the help of a GPS. If this is new to you the front seat passenger is the one in charge of navigation and directing the vehicle. The driver's role is more focused on physically getting the vehicle safely along the route.

Mounted land nav conditions can include interstate movement, secondary roads, forest service trails, city streets, goat trails, and off-road mounted movement. Always remember you get the bad with the good when mounted. You are probably in an unarmored vehicle that attracts a lot of attention - and you also have all the fueling and maintenance headaches. But the mobility advantages of being mounted (even on a simple mountain bike) can be worth it.

Navigating on fast single rider vehicles presents challenges to go along with all the advantages. GPS makes this a lot easier, but you still need to memorize and mentally drive / visualize your routes beforehand.

Don't let yourself get too focused on the vehicle, mounted land nav success is all about **anticipating the route and staying ahead of the decision cycle**. You must be proactive vs reactionary. There is no room for error or last second "turn here!" directions to your driver. If you are operating on a single person vehicle the challenges are even greater, there is no one to assist with the navigation process.

185

When planning your route, the effects of terrain and trafficability for your type of vehicle must be determined (OAKOC). One of the least desired events is being forced to become a dismounted navigator again because you got your vehicle stuck or had a rollover. A very real possibility during riots or civil unrest are roadblocks and debris placement; planning for bypasses around templated obstacles is part of your movement planning. Expand and adjust the terrain analysis techniques you know how to use for dismounted land nav. Apply all the fundamentals to keep you out of trouble.

Having the mobility advantage can make it faster / easier if you get disoriented to move to a point where you can reorient yourself. But it can also be a detriment if you are in restrictive terrain with limited mobility corridors. Restrictive terrain may not always be swamps and cliffs; based on conditions a dense built up area (urban) can become severely restricted terrain. If you are inside a city during a social crisis, you may find primary routes completely cut off and lateral routes clogged by coordinated protesters. This will quickly turn the vehicle advantages into liabilities.

Urban canyons are a challenge for navigation. Not only do they canalize your movement and hide lateral egress routes, they can block GPS signals from getting through.

186

The key to maneuvering a vehicle is keeping your options open; don't get you, your team, family, or your vehicle in a position where you have no way out. Retain your freedom of maneuver and plan your routes in a way that does not narrow your options to only one.

Vehicles have mobility limitations on terrain that would be no problem if you were dismounted. When determining a route do not overestimate the capabilities of your off road vehicle(s). I managed to mire a YZ250 motorcycle in a muddy creek bed that took three dudes and hand winch to get it out. A series of bad terrain can completely stop a vehicle and turn simple navigation into a recovery exercise...or worse if there is a hostile entity.

Mounted nav is not just about getting to a destination. It is about keeping your vehicle mission capable and your options open. Bad route decisions can have significant and lasting consequences during a crisis.

VEHICLE MAP SETS

Various scale maps are a must for your vehicle and GHB. From a large scale 1:24k top map all the way to small scale road atlases you should consider having options. Use the recommendations from Chapter 1 to help you choose your map sets and tailor yours to the type of vehicle and area.

The Delorme® state topo map books are outstanding for mounted land navigation. They are a good balance of scale and convenience. Have a current version of your state and any border state you venture to on a regular basis. I used these for years when duck hunting in pre smartphone years. They never let me down. These maps provide excellent details of forest trails, put ins, and side streets and roads that do not appear on state roadmaps. The scales vary on these depending on the state, so be aware they may not be the same for all your areas and it is becoming exceedingly difficult to find protractors / scales that work with them.

Have a system and a map set in your vehicle / get home bag. A well thought out set of maps as discussed in Chapter 1 will cover your requirements. You will need more than a basic road atlas.

MOUNTED TERRAIN ASSOCIATION

Terrain association uses the same concepts as when dismounted, it just moves so much faster. You will plan vehicle routes to move through the terrain while still accounting for all the METT-TC and OAKOC factors. You will most likely be using a wheeled vehicle that is fast and mobile, but even a 4x4 with off road tires has significant limitations off trail. A dirt bike or ATV has advantages off road and can open options for you, but there are always tradeoffs (cargo, passengers, speed, noise). The reality of this is you will most likely not be navigating a vehicle across a huge maneuver area devoid of civilization, trails, and human built features. There is however the strong possibility of GPS system disruptions. Disruptions that may be coupled with a lot of chaos and route closures that force you to bypass by using non-traditional routes. As you read through this section keep these conditions in mind.

Using terrain association on a vehicle will seem familiar to you; it is a blend of the dismounted navigation skills coupled with the common driving skills you are used to. Moving from terrain feature to terrain feature is the same process and uses similar mental checkpoints from a familiar drive through your town or city. For terrain association you are guiding on map features (ridge, hill, stream) vs the daily drive things such as stores, restaurants, and intersections.

Remember during both your planning and execution your route must be capable of sustaining the travel of the vehicle (or vehicles), should be relatively direct, and should be easy to follow. The terrain along the flanks (sides) of your route will take on more significance in a social crisis. All those laterals and bypasses we talked about must be in a constant state of change. A vehicle commander (navigator) must always be looking for options.

In a typical mounted move, the navigator determines current location, the location of the objective, annotates the position of both on his map, and then selects a route between the two. After assessing the terrain you will adjust your route (apply METT-TC and OAKOC from Chapter 2). Nothing new here.

Consider Ease of Movement. Use the easiest possible route and bypass difficult terrain. Remember that a difficult route is harder to follow, is noisier, causes more wear and tear (and possible recovery problems), and takes more time. Try to select an axis or corridor instead of a specific route. Make sure you have enough maneuver room for the vehicles you are using. For example, you won't be able to take a UTV down a single track trail.

Use Features as Checkpoints. These checkpoints must be easily recognizable in the light and weather conditions and at the speed at which you will move. You should be able to find a feature from your location that can be recognized from almost anywhere and used as a guide. The best checkpoints are linear features that cross your route. Perennial streams, rivers, hardball roads, ridges, valleys, and railroads are all suitable checkpoints for mounted nav.

Follow Terrain Features. Movement and navigation along a valley floor or near (not necessarily on) the crest of a ridgeline is easiest. Unfortunately, vehicles are best suited for movement along natural lines of drift and can push us into choosing exposed routes. Trying to stay hidden and out of sight is an obvious downside of mounted navigation. However, you might be surprised at how stealthy a vehicle (even a 68 ton tank) can be when in the right hands. Terrain driving is the technique of skirting high ground and using terrain masking (keeping terrain between you and any threats) when you can on your routes.

Determine Directions. Break the route down into smaller segments and determine the rough directions that will be followed. You do not need to use a specific azimuth; just use the cardinals or a general azimuth to orient the route. Even using the on-board vehicle compass that are standard on most modern automobiles is a good means to stay oriented. Locate changes of direction, if any, at the checkpoints you chose along your route.

Determine Distance. Get the total route distance to be traveled and the approximate distance between checkpoints. Use the vehicle odometer to help keep track of distance traveled.

Visualize and rehearse the route. Try to imagine what the route is like and remember it. Rehearse the route on a terrain model if this is a cross country route. Small scale road maps laid out on the ground are a basic rehearsal technique you can use to help think through the problem.

Set yourself up for success. Same as with dismounted route planning be sure to build in those confidence checkpoints. Be deliberate about choosing terrain and features when you know there is a difficult stretch to navigate through. Study and mentally rehearse the route to determine where errors are most likely to occur and plan checkpoints and short legs to help you maintain your route.

Dead Reckoning for mounted land nav. The components needed to dead reckon (point navigation) mounted are the same as dismounted; you need a distance and a direction. The chances of having to dead reckon from a vehicle are slim, and it is used only as a last ditch effort to navigate if you are mounted. You will most likely be using terrain association if you aren't using GPS, but I want to go over some highlights of mounted dead reckoning just in case you need it in the middle of a featureless landscape. Dead reckoning takes away the speed advantage a vehicle provides so it should be used as a last resort.

Due to the magnetic interference of a vehicle's metal mass (even the metal in a mountain bike) there is no accurate method of determining the direction from a moving vehicle other than a GPS based compass. You will need to dismount and move out of the vehicle's field to shoot an accurate azimuth with a standard compass. Dismounting and finding distant steering marks or getting a bearing are on the list of potential actions while navigating mounted without a GPS. We won't go any deeper on this vehicle dead reckoning piece for now. Unless you are on the featureless plains or flat desert it is something that has limited use.

MOUNTED BYPASS

Navigating the planned route is the easy part of mounted land navigation. The expert level of this is being able to quickly find and execute a bypass when you encounter an issue along your route. A large part of keeping yourself safe and avoiding roadblocks, protests and mobs is knowing how to bypass the criminal activity. Teach yourself how to do this quickly without using electronics (GPS). Learn how to immediately reroute yourself around a blocked street or trouble area. Do not count on road signs being present or correct, bad actors are known to remove or even reposition road signs to create confusion.

In a widespread crisis you may not be able to trust road signs. During the Battle of the Bulge the Germans would either destroy signs or repost incorrect signage to create confusion. Expect the same by other modern bad actors. (photo permission of Truman Presidential library)

COMMON INFRASTRUCTURE TERMS

We don't have to be full on Armageddon to see value in mounted navigation skills. Even during a widespread crisis, we may have infrastructure and signage that can help us navigate. We shouldn't rely solely on signs and the naming conventions listed below but knowing the terminology and the road numbering systems (in the US) can help you make navigation decisions.

This section is not meant to insult anyone's experience or intelligence, many of you are aware of these terms and numbering nuances already. Based on the pre-surveys we did for this manual we found that there are enough folks unfamiliar with naming conventions that is warrants a couple of pages.

Modern city planners have played fast and loose with naming conventions, so beware some of these are not set in stone (especially street vs road). But they are still of use for mounted land nav general knowledge and reference.

Street - exist in built up areas (cities, towns, etc) and are lined with structures (buildings) on either side. Unlike roads, streets allow travel in/through a town, not to connect two townships together. By definition streets are always paved.

Road - roads connect two points (two towns or two communities). Roads are typically found in more rural areas, but roads are also in cities as well.

Avenue - runs perpendicular to streets, so a city grid will probably have streets and avenues cross hatched.

Boulevard - a wide street with several lanes and a median.

Cul-de-sac or *Court* – end of a residential street with a dead end or short loop; one way in one way out.

Drive - a scenic route around a mountain or natural resource. This is a good example of the planners not abiding by their own rules. Plenty of named "Drives" inside of American suburban neighborhoods.

Alley - a narrow side road, sometimes dead ends

Lane - a narrow road or part of a larger street

Place – neighborhood street that leads to a dead end

Way - a side street

INTERSTATE NUMBERING SYSTEM

The US Interstate numbers have a system that can assist you with navigation. North/South interstates are odd numbers, the East/West ones are even. The longer interstates (or "majors") that run east/west end in zero, and north/south majors end in 5.

➤ Interstate 5 or "I-5"
➤ Odd number so you know it runs generally North/South.
➤ It is a low number, so you know it is near the west coast.
➤ It ends in a 5 so you know it is a "major"

The numbers are listed in numerical order from west to east and south to north. For example, I-90 is the most northern east/west interstate and runs from Seattle to Boston; the southernmost east-to-west US interstate is I-10.

➤ Interstate 90 or "I-90"
➤ Even number so you know it runs generally East/West.
➤ It is a high number, so you know it cuts across the northern US.
➤ It ends in a 0 so you know it is a "major"

"Big picture" view of how the numbering system works for the major interstates. North/South are odd and end in 5 and East/West are even and end in 0.

An interstate with 3 digits is an auxiliary interstate and is a branch of the main highway. These generally occur around a large city. They can be spurs, bypasses, or beltways. An even starting digit (I-495) means that the auxiliary road connects with an interstate at both ends, or as a loop. In general, most that start with a 4XX are loops. For example, I-495 is a beltway around DC and intersects with I-95.

Three digits are Auxilary Interstates and are always in the vicinity of large urban areas.

A three digit Auxilary interstate with an odd starting digit (eg I-380) means that it is connected to it's parent interstate only at one end.

Three digit with an odd first number are spurs and will connect only at one end with the parent interstate.

A bypass route may go around a city or may run through it. Typical 3-digit Interstate Highway bypasses usually have both ends joined with another interstate. Bypass routes are preceded by an even number in the first digit.

Three digit with an even first number are bypasses. These can also be beltways.

Backward planning. Backward planning is the process of starting at the end of your route / destination and then planning the events "backward". It is also called "reverse planning." Reverse planning will allow you to visualize the entire movement across time. If you must be at your destination by a specific time backwards planning will let you know when to leave. Route planning tools such as GPS and mobile navigation apps do this for you automatically. You will use this process no matter your means of transportation (mounted or dismounted). The speeds and complexities of mounted navigation require additional planning time and attention to detail, the backwards planning model will help you visualize and plan the movement.

   Some reverse planning considerations. Understand your "endstate" or the specifics of where you want to be at the end of the movement and the condition you want to be in. For example, your endstate may be to arrive at the destination with at least a half tank of fuel left in your vehicle. The implied task for this requirement is you must plan a fuel stop or a refuel on the move (ROM) using on board fuel cans. You may have to account for water resupply along the way; this could involve adjusting your route to pass by a water source. These details seem silly in normal conditions, but if things fall apart the details will be what keep you on track. Practice these skills in normal times, you won't have time to learn on the fly during a crisis.

Mounted land nav starts to cross into tactics and the challenge of self-supporting logistics, you should apply some of the troop leading procedure (TLP) concepts to any navigation training. Capture your timeline in a way so you can visualize it from start to finish, it is part of your route planning. This backwards planning tool can be whatever you like as long as it gives you the information you need.

Timeline example. The timeline on the next page is something you can use to visualize any movement. Note the planning time does not equal the total amount of hours before the first movement (when you leave your assembly area, campsite etc). You must account for time that is used for other things like sleep, maintenance, mission planning, and even personal hygiene. This applies to vehicle or dismounted movement, the principles are the same. There is not a set way to do this, adapt and use what works for you.

Have a tool that gives you a visual reference or "picture" of your timeline to help reveal issues with your plan and give you a chance to correct them. You can be as detailed as you want with this, over time you will learn what is important to include on your timeline and what is not. Please note this is not a full reverse planning sequence, for tactical missions use the CM-2 Recon manual as a reference for planning and timeline development. The blue text in the example on the next page depicts items you would write in yourself. We also included a blank version of this timeline for your use in Appendix D of this manual.

Sample Reverse Planning Timeline (simplified)

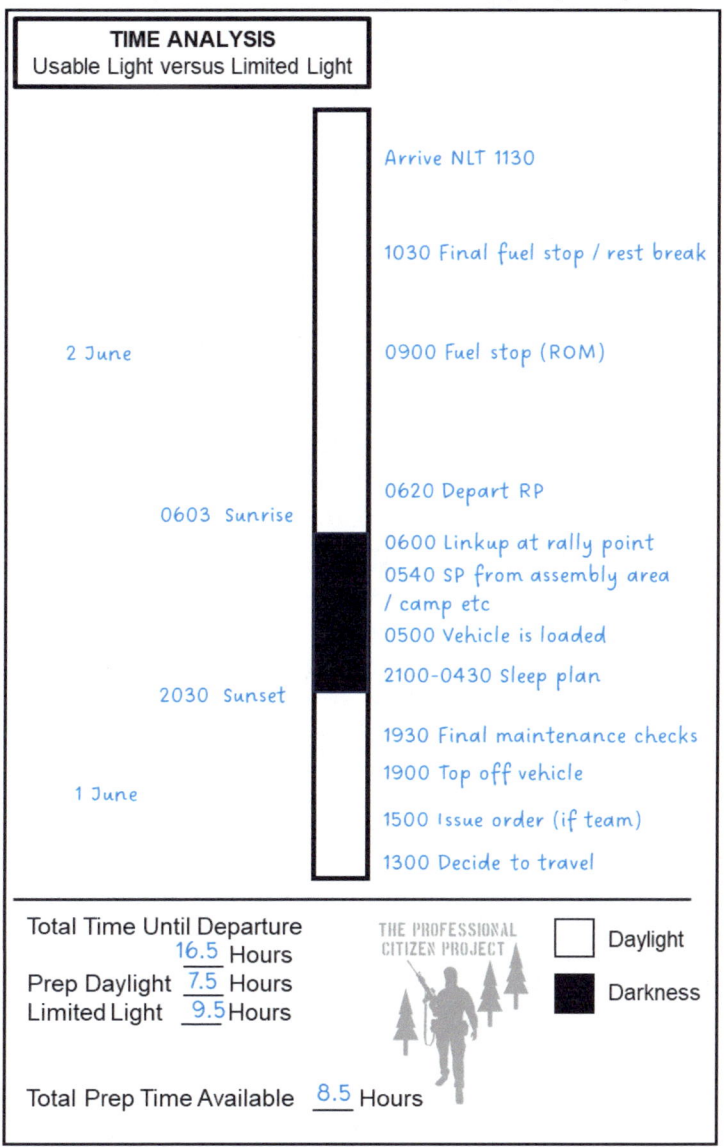

TIME ANALYSIS
Usable Light versus Limited Light

Arrive NLT 1130

1030 Final fuel stop / rest break

2 June

0900 Fuel stop (ROM)

0620 Depart RP

0603 Sunrise

0600 Linkup at rally point
0540 SP from assembly area / camp etc
0500 Vehicle is loaded

2100-0430 Sleep plan

2030 Sunset

1930 Final maintenance checks

1900 Top off vehicle

1 June

1500 Issue order (if team)

1300 Decide to travel

Total Time Until Departure
__16.5__ Hours
Prep Daylight __7.5__ Hours
Limited Light __9.5__ Hours

THE PROFESSIONAL
CITIZEN PROJECT

☐ Daylight
■ Darkness

Total Prep Time Available __8.5__ Hours

CONCLUSION

There are hundreds of volumes written about the land nav subject, thank you for choosing this one and trusting us to provide the content. Remember this manual is not all-encompassing, there are primitive land nav techniques and extremely technical methods that we did not address in here. Being able to choose from a comprehensive set of skills and knowing which apply best is the position we want you to be in. Learn the hard skills inside here and then get out and apply them. Take a class, join a local group and have fun doing it. There are sport and enthusiast events that can help you practice your navigation skills; geo caching, and orienteering are a couple of good ways to reinforce your skills. Good luck and stay found.

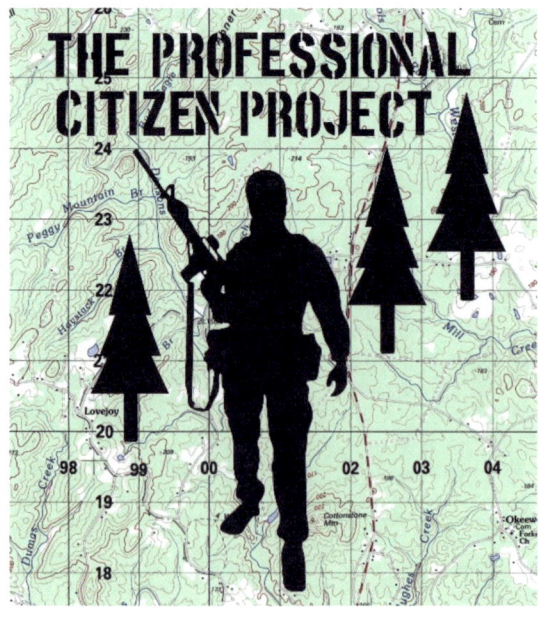

APPENDIX A
Navigation Gear Checklist

This is in addition to your mission PCC/PCI Checklist, recommend incorporating this into your SOPs.

Map
- ☐ Covers your anticipated area
- ☐ Waterproofing
- ☐ Current declination verified

Compass
- ☐ Primary and Alternate compasses on hand
- ☐ Inspected and verified

Writing Gear
- ☐ 2ea pencils (at least one mechanical .5 to .9) functioning with extra lead
- ☐ Waterproof notebook (RiteintheRain® recommended)

Protractor
- ☐ Commercial versions, verify scale for map, carry a backup also
- ☐ Straight edge (can be large protractor)

Miscellaneous
- ☐ Pace beads
- ☐ Monocular 8 or 10x (to check routes and danger areas as you move)
- ☐ Rangefinder
- ☐ Power (battery packs, spare lithium batteries)
- ☐ _____
- ☐ _____
- ☐ _____

APPENDIX B
USGS Symbols
These are directly from the USGS

BATHYMETRIC FEATURES

Area exposed at mean low tide; sounding datum line***	
Channel***	
Sunken rock***	+

BOUNDARIES

National	
State or territorial	
County or equivalent	
Civil township or equivalent	
Incorporated city or equivalent	
Federally administered park, reservation, or monument (external)	
Federally administered park, reservation, or monument (internal)	
State forest, park, reservation, or monument and large county park	
Forest Service administrative area*	
Forest Service ranger district*	
National Forest System land status, Forest Service lands*	
National Forest System land status, non-Forest Service lands*	
Small park (county or city)	

BUILDINGS AND RELATED FEATURES

Building	
School; house of worship	
Athletic field	
Built-up area	
Forest headquarters*	
Ranger district office*	
Guard station or work center*	
Racetrack or raceway	
Airport, paved landing strip, runway, taxiway, or apron	
Unpaved landing strip	
Well (other than water), windmill or wind generator	
Tanks	
Covered reservoir	
Gaging station	
Located or landmark object (feature as labeled)	
Boat ramp or boat access*	
Roadside park or rest area	
Picnic area	
Campground	
Winter recreation area*	
Cemetery	Cem

COASTAL FEATURES

Foreshore flat	
Coral or rock reef	Reef
Rock, bare or awash; dangerous to navigation	
Group of rocks, bare or awash	
Exposed wreck	
Depth curve; sounding	
Breakwater, pier, jetty, or wharf	
Seawall	
Oil or gas well; platform	

CONTOURS

Topographic

Index	
Approximate or indefinite	
Intermediate	
Approximate or indefinite	
Supplementary	
Depression	
Cut	
Fill	
Continental divide	

Bathymetric

Index***	
Intermediate***	
Index primary***	
Primary***	
Supplementary***	

CONTROL DATA AND MONUMENTS

Principal point**	3-20
U.S. mineral or location monument	USMM 438
River mileage marker	Mile 69

Boundary monument

Third-order or better elevation, with tablet	BM 9134 BM 277
Third-order or better elevation, recoverable mark, no tablet	5628
With number and elevation	67 4567

Horizontal control

Third-order or better, permanent mark	Neace Neace
With third-order or better elevation	BM 52 Pike BM393
With checked spot elevation	1012
Coincident with found section corner	Cactus Cactus
Unmonumented**	+

204

CONTROL DATA AND MONUMENTS – *continued*

Vertical control

Third-order or better elevation, with tablet	BM ✕ 5280
Third-order or better elevation, recoverable mark, no tablet	✕ 528
Bench mark coincident with found section corner	BM + 5280
Spot elevation	✕ 7523

GLACIERS AND PERMANENT SNOWFIELDS

Contours and limits	
Formlines	
Glacial advance	
Glacial retreat	

LAND SURVEYS

Public land survey system

Range or Township line	
Location approximate	
Location doubtful	
Protracted	
Protracted (AK 1:63,360-scale)	
Range or Township labels	R1E T2N R3W T4S
Section line	
Location approximate	
Location doubtful	
Protracted	
Protracted (AK 1:63,360-scale)	
Section numbers	1 - 36 1 - 36
Found section corner	
Found closing corner	
Witness corner	WC
Meander corner	MC
Weak corner*	

Other land surveys

Range or Township line	
Section line	
Land grant, mining claim, donation land claim, or tract	
Land grant, homestead, mineral, or other special survey monument	□
Fence or field lines	

MARINE SHORELINES

Shoreline	
Apparent (edge of vegetation)***	
Indefinite or unsurveyed	

MINES AND CAVES

Quarry or open pit mine	✕
Gravel, sand, clay, or borrow pit	✕
Mine tunnel or cave entrance	⊰
Mine shaft	▫
Prospect	✕
Tailings	(Tailings)
Mine dump	
Former disposal site or mine	

PROJECTION AND GRIDS

Neatline	39°15′ 90°37′30″
Graticule tick	55′
Graticule intersection	
Datum shift tick	

State plane coordinate systems

Primary zone tick	640 000 FEET
Secondary zone tick	247 500 METERS
Tertiary zone tick	260 000 FEET
Quaternary zone tick	98 500 METERS
Quintary zone tick	320 000 FEET

Universal transverse mercator grid

UTM grid (full grid)	273
UTM grid ticks*	269

RAILROADS AND RELATED FEATURES

Standard gauge railroad, single track	
Standard gauge railroad, multiple track	
Narrow gauge railroad, single track	
Narrow gauge railroad, multiple track	
Railroad siding	
Railroad in highway	
Railroad in road	
Railroad in light duty road*	
Railroad underpass; overpass	
Railroad bridge; drawbridge	
Railroad tunnel	
Railroad yard	
Railroad turntable; roundhouse	

RIVERS, LAKES, AND CANALS

Perennial stream	
Perennial river	
Intermittent stream	
Intermittent river	
Disappearing stream	
Falls, small	
Falls, large	
Rapids, small	
Rapids, large	
Masonry dam	
Dam with lock	
Dam carrying road	

205

RIVERS, LAKES, AND CANALS – *continued*

Perennial lake/pond	
Intermittent lake/pond	
Dry lake/pond	(Dry Lake)
Narrow wash	
Wide wash	< Wash
Canal, flume, or aqueduct with lock	
Elevated aqueduct, flume, or conduit	
Aqueduct tunnel	
Water well, geyser, fumarole, or mud pot	o o
Spring or seep	

ROADS AND RELATED FEATURES

Please note: Roads on Provisional-edition maps are not classified as primary, secondary, or light duty. These roads are all classified as improved roads and are symbolized the same as light duty roads.

Primary highway	
Secondary highway	
Light duty road	
Light duty road, paved*	
Light duty road, gravel*	
Light duty road, dirt*	
Light duty road, unspecified*	
Unimproved road	
Unimproved road*	
4WD road	
4WD road*	
Trail	
Highway or road with median strip	
Highway or road under construction	Under Const
Highway or road underpass; overpass	
Highway or road bridge; drawbridge	
Highway or road tunnel	
Road block, berm, or barrier*	
Gate on road*	
Trailhead*	

* USGS-USDA Forest Service Single-Edition Quadrangle maps only.
In August 1993, the U.S. Geological Survey and the U.S. Department of Agriculture's Forest Service signed an Interagency Agreement to begin a single-edition joint mapping program. This agreement established the coordination for producing and maintaining single-edition primary series topographic maps for quadrangles containing National Forest System lands. The joint mapping program eliminates duplication of effort by the agencies and results in a more frequent revision cycle for quadrangles containing National Forests. Maps are revised on the basis of jointly developed standards and contain normal features mapped by the USGS, as well as additional features required for efficient management of National Forest System lands. Single-edition maps look slightly different but meet the content, accuracy, and quality criteria of other USGS products.

SUBMERGED AREAS AND BOGS

Marsh or swamp	
Submerged marsh or swamp	
Wooded marsh or swamp	
Submerged wooded marsh or swamp	
Land subject to inundation	Max Pool 4.31

SURFACE FEATURES

Levee	Levee
Sand or mud	Sand
Disturbed surface	
Gravel beach or glacial moraine	Gravel
Tailings pond	Tailings Pond

TRANSMISSION LINES AND PIPELINES

Power transmission line; pole; tower	
Telephone line	Telephone
Aboveground pipeline	
Underground pipeline	Pipeline

VEGETATION

Woodland	
Shrubland	
Orchard	
Vineyard	
Mangrove	Mangrove

** Provisional-Edition maps only.
Provisional-edition maps were established to expedite completion of the remaining large-scale topographic quadrangles of the conterminous United States. They contain essentially the same level of information as the standard series maps. This series can be easily recognized by the title "Provisional Edition" in the lower right-hand corner.

*** Topographic Bathymetric maps only.

Topographic Map Information
For more information about topographic maps produced by the USGS, please call:
1-888-ASK-USGS or visit us at http://ask.usgs.gov/

APPENDIX C
Map Modifications
This is a recommended example of mods that worked for me in the past; adjust to fit your requirements.

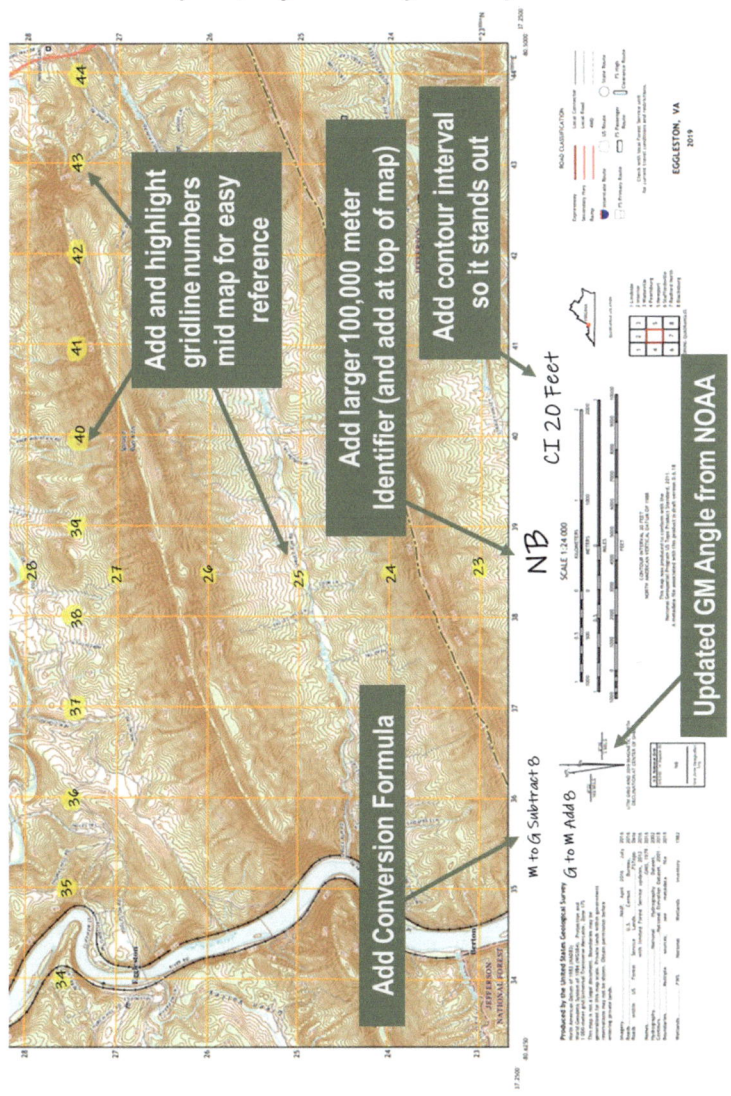

Add and highlight gridline numbers mid map for easy reference

Add larger 100,000 meter Identifier (and add at top of map)

Add contour interval so it stands out

CI 20 Feet

Updated GM Angle from NOAA

Add Conversion Formula

M to G Subtract B   G to M Add B

APPENDIX D
Reverse Planning Timeline (blank)

Reverse Planning Timeline

TIME ANALYSIS
Usable Light versus Limited Light

Total Time Until Departure
_____ Hours
Prep Daylight _____ Hours
Limited Light _____ Hours

Total Prep Time Available _____ Hours

THE PROFESSIONAL
CITIZEN PROJECT

☐ Daylight
■ Darkness

APPENDIX E
100m Pace Count Record
(Add your own conditions to meet your requirements but don't overcomplicate it)

Pace Count	Day/ Night	Conditions
	Day	Open Flat Field No Gear
	Day	Open Flat Field w/Load
	Day	Over Varied Terrain w/Load
	Night	Over Varied Terrain w/Load
	Day	Road or Trail w/Load at a Shuffle (Running Pace Count)

References

Citizen Manual 1 (CM-1) Individual Tactical Skills, Jack Morris, 1 June 2023

Citizen Manual 2 (CM-2) Reconnaissance, Jack Morris 2023

Citizen Manual 9 (CM-9) Adverse Conditions and Environments, Jay Pallardy, 2023

FM 3-25.26 Map Reading and Land Navigation

ATP 2-01.3. *Intelligence Preparation of the Battlefield/Battlespace,* 10 November 2014

ATP 3-20.98. *Reconnaissance Platoon,* 5 April 2013

FM 3-98. *Reconnaissance and Security Operations,* 1 July 2015

ATP 3-50.21, US Army, Survival

Jack Morris is a retired US Army officer with 25 years of service as both an enlisted soldier and commissioned officer. Jack has led small recon teams, squads, armor and infantry platoons, and companies. Jack is a leader development expert with years of experience leading, training, developing, and mentoring leaders for the Army. He is an experienced doctrine writer (leadership and maneuver warfare) and has years of accumulated time at maneuver battalions, brigades, and heavy divisions as an operations officer leading brigade and division level command posts. Jack is instrumental in the development of the Pro Citizen series of manuals and leads the Professional Citizen Project team's efforts to build and publish the Pro Citizen series references.

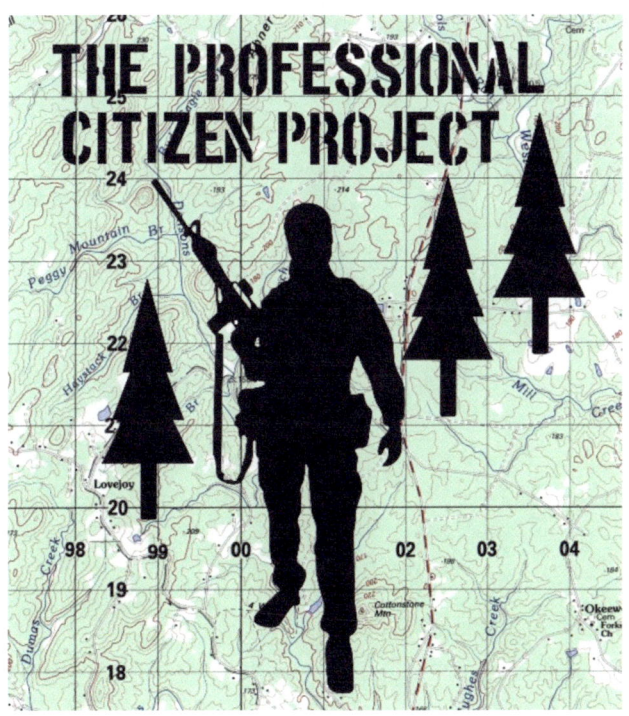

Additional Citizen Manuals from the Professional Citizen series of references for your personal library are available from us at The Professional Citizen Project at *TPCProject.com* or through Amazon. We are always working on new manuals for the community. Follow us on our IG page and sign up for our weekly newsletter to receive exclusive content, new releases, and our weekly discount codes.

CM-1 Individual Tactical Skills by Jack Morris

CM-2 Reconnaissance by Jack Morris

CM-3 The AR Pattern Rifle and Carbine by Mike Lewis

CM-4 Teenage Prepper by Cody and Jack Morris

CM-5 Land Navigation by Jack Morris

CM-9 Adverse Environments and Conditions (ACE) by Jay Pallardy

Made in United States
Troutdale, OR
03/08/2025

29597586R00122